Mammal Ecology

TERTIARY LEVEL BIOLOGY

A series covering selected areas of biology at advanced undergraduate level. While designed specifically for course options at this level within Universities and Polytechnics, the series will be of great value to specialists and research workers in other fields who require a knowledge of the essentials of a subject.

Titles in the series:

Experimentation in Biology	Ridgman
Methods in Experimental Biology	Ralph
Visceral Muscle	Huddart and Hunt
Biological Membranes	Harrison and Lunt
Comparative Immunobiology	Manning and Turner
Water and Plants	Meidner and Sheriff
Biology of Nematodes	Croll and Matthews
An Introduction to Biological Rhythms	Saunders
Biology of Ageing	Lamb
Biology of Reproduction	Hogarth
An Introduction to Marine Science	Meadows and Campbell
Biology of Fresh Waters	Maitland
An Introduction to Developmental Biology	Ede
Physiology of Parasites	Chappell
Neurosecretion	Maddrell and Nordmann
Biology of Communication	Lewis and Gower
Population Genetics	Gale
Structure and Biochemistry of Cell Organelles	Reid
Developmental Microbiology	Peberdy
Genetics of Microbes	Bainbridge
Biological Functions of Carbohydrates	Candy
Endocrinology	Goldsworthy, Robinson and Mordue
The Estuarine Ecosystem	McLusky
Animal Osmoregulation	Rankin and Davenport
Molecular Enzymology	Wharton and Eisenthal
Environmental Microbiology	Grant and Long
The Genetic Basis of Development	Stewart and Hunt
Locomotion of Animals	Alexander
Animal Energetics	Brafield and Llewellyn
Biology of Reptiles	Spellerberg
Biology of Fishes	Bone and Marshall

Mammal Ecology

M. J. DELANY, M.Sc., D.Sc.

Professor of Environmental Science
University of Bradford

Blackie

Glasgow and London
Distributed in the USA by
Chapman and Hall
New York

Blackie & Son Limited,
Bishopbriggs, Glasgow G64 2NZ

Furnival House, 14–18 High Holborn, London WC1V 6BX

Distributed in the USA by
Chapman and Hall
in association with Methuen, Inc.
733 Third Avenue, New York, N.Y. 10017

599.05

© 1982 Blackie & Son Ltd
First published 1982

British Library Cataloguing in Publication Data

Delany, M. J.
  Mammal ecology.—(Tertiary level biology)
  1. Mammals
  I. Title   II. Series
  599   QL703

  ISBN 0-216-91310-1
  ISBN 0-216-91309-8 Pbk

Library of Congress Cataloging in Publication Data

Delany, M. J.
  Mammal ecology.

  (Tertiary level biology)
  Bibliography: p.
  Includes index.
  1. Mammals—Ecology.   I. Title.   II. Series.
  QL703.D44 1982   599.05   82-4592
  ISBN 0-412-00081-4   AACR2
  ISBN 0-412-00091-1 (pbk.)

Filmset by Advanced Filmsetters (Glasgow) Ltd

Printed in Great Britain by
Thomson Litho Ltd, East Kilbride, Scotland

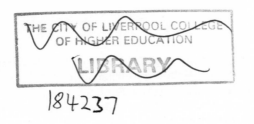
Preface

The mammals, with their multitudinous range of structure and function from the massive whales to the smallest shrews, are a well-known and easily recognized group of animals. They have attracted man's interest for thousands of years, with many species adapting in various ways to live in close association with him either under domestication or as notorious pests. Although much has been learned about the domestic species, it is probably true that knowledge of the ecology of wild species, which form the vast majority, was extremely limited as recently as three decades ago. Since about 1950, there has been a considerable and continuing expansion of research into these animals all over the world, which has produced a new and rich scientific literature. The present account brings together a representative selection of this material within the confines of a small volume, to provide a balanced picture of the broad lines of progress achieved to date. Throughout the book, the reader is made aware of some of the limits of present knowledge and the opportunities for further research. An extensive and comprehensive bibliography to original sources and advanced specialist texts has been included, to allow individual points to be followed up in detail.

Progress in recent years can give no cause for belief that little remains to be discovered. The emergence of new technologies, from sophisticated computers to microchips, offers opportunities for types of investigation inconceivable only a few years ago. Furthermore, in much of the world, notably in many parts of the tropics, knowledge of the ecology of the mammalian inhabitants frequently ranges from the modest to the negligible with, as a result, opportunities offered for the most fundamental studies.

This account is designed primarily to expand the student's and serious naturalist's knowledge of mammals. As such, it is expected to be of use in university and college courses in mammalogy, and as an adjunct to courses in such fields as vertebrate biology, ecology and behaviour. It will also be of value to individuals who deal with mammals in the course of their work—game biologists, park wardens, agriculturalists, pest control officers and veterinarians, for example. There is no assumption that the reader will have a detailed knowledge of ecology nor of mammal systematics, although an acquaintance with the main characteristics of the major orders would be helpful.

Some of the information contained in this book has been kindly provided by colleagues. I would like to thank Mr V. P. W. Lowe for the data on red deer on the Island of Rhum (Fig. 5.7) and Professor Z. Pucek and Dr Z. A. Krasinski for the figures on the bison populations in Bialowieza Forest, Poland (Fig. 5.12).

M.J.D.

Contents

CHAPTER ONE

THE MAMMALS

The mammals are an easily recognizable and readily definable group of animals. They are the only animals that possess hair and, together with the birds, are the only group having a constant body temperature, i.e. are homeothermic. This makes them more independent of environmental conditions, notably high and low temperatures, than is the case for those animals unable to regulate their body temperature (poikilotherms, which comprise the vast majority of species in the animal kingdom). The reproduction of mammals is unique, as they are the only animals that produce milk. Fertilization is internal. The monotremes are exceptional among the mammals in producing shelled eggs, which they incubate. The marsupials have an abbreviated development in the uterus, while in the remaining orders (Eutheria or placentals) there is a protracted development. A further characteristic of the mammals is an elaborate dentition comprising several different types of teeth (incisors, canines, premolars and molars). This, together with the internal nares opening immediately anterior to the pharynx, has facilitated retention of food in the mouth and its mastication. Dentitions are adapted for utilizing a wide range of different foods, including fruits, leaves and shoots, insects and larger animals, and various combinations of these. Structures associated with the perception of the senses of sight, sound, smell and taste are often extremely well developed, as is the brain, their co-ordinating centre. Its enlargement also results in the facility of learning from past actions and the appearance in some mammals of intelligence, or the ability to interrelate different previous experiences.

The first mammals, which appeared during the Triassic about 200 million years ago, were small mouse-sized animals which had evolved from Synapsid reptiles. As a group mammals display greater functional efficiency than their antecedents and have not, like the birds, evolved into just one mode of life. It was not until the Palaeocene and Eocene,

1

approximately 60–70 million years ago, that most of the modern orders, as we know them today, appeared. About this time, there was a great radiation of placental mammals along several diverse lines, notably from a common, basic, insectivorous stock. The pattern of this evolution was greatly influenced by habitat and feeding adaptations. Thus, some stocks came to utilize trees, sea and sky to become primates, whales and bats respectively, while among the terrestrial forms, some became herbivorous and others carnivorous to give rise subsequently to groups such as the rodents, ungulates and carnivores. By the Oligocene, many of the families present were those in existence today and the major evolution of the class was complete. There are 21 orders of living mammals (Table 1.1).

Although an extremely successful group, the mammals possess comparatively few species (a total of just over 4000). This is quite modest compared to an estimated 750 000 + species of insects, 100 000 molluscs, 50 000 Protozoa and even 6500 echinoderms. Included in the 31 000 species of vertebrates are 11 500 fish and 8500 birds. The small number of species of mammals is probably related to their relatively large size. This condition is forced upon them through the maintenance of a constant body temperature, as decreasing size results in larger surface area to volume ratios and thereby proportionately larger surfaces for heat loss. The consequent absence of mammals the size of insects denies them the opportunity to exploit specialized micro-habitats and so reduces the numbers of niches available to them. They do, nevertheless, show an astonishing range of adaptation and specialization, with such diverse trophic habits as, for example, some bats that live exclusively on nectar and others on blood, the koala (*Phascolarctos*) that eats eucalyptus leaves, the giant panda (*Ailuropoda*) that consumes only bamboo shoots and several species of anteater (Tachyglossidae, Myrmecophagidae). There are also many generalist feeders whose requirements are less specific, but who nevertheless are as well adapted to exploiting their food resource as the specialists. It is thus evident that the relative paucity of species does not result in a group of uniform ecological adaptation and habit.

Table 1.1   The orders and families of mammals (after Corbet and Hill, 1980). The number of species is given in brackets. Abbreviations: Af., Africa; Am., America; Arc., Arctic; As., Asia; Aus., Australia; C., Central; E., East; Eur., Eurasia; Hem., Hemisphere; Mad., Madagascar; N., North; New G., New Guinea; New Z., New Zealand; O. W., Old World; S., South; W., West; Wo., Worldwide.

ORDER MONOTREMATA   Monotremes (3)
   Tachyglossidae, spiny anteaters, Aus., New G. (2).
   Ornithorhynchidae, platypus, Aus. (1).

ORDER MARSUPIALIA   Marsupials (254)
  Didelphidae, American opossums, Am. (73).
  Microbiotheriidae, colocolos, S. Am. (1).
  Caenolestidae, shrew-opossums, S. Am. (7).
  Dasyuridae, marsupial mice, Aus., New G. (49).
  Myrmecobiidae, numbat, Aus. (1).
  Thylacinidae, thylacine, Aus. (1).
  Notoryctidae, marsupial mole, Aus. (1).
  Peramelidae, bandicoots, Aus., New G. (17).
  Thylacomyidae, rabbit-bandicoots, Aus. (2).
  Phalangeridae, phalangers, Aus., New G. (11).
  Burramyidae, pygmy possums, Aus., New G. (7).
  Petauridae, gliding phalangers, Aus., New G. (22).
  Macropodidae, kangaroos, wallabies, Aus., New G. (57).
  Phascolarctidae, koala, Aus. (1).
  Vombatidae, wombats, Aus. (3).
  Tarsipedidae, honey possum, Aus. (1).

ORDER EDENTATA   Edentates (29)
  Myrmecophagidae, American anteater, S. C. Am. (4).
  Bradypodidae, sloths, S. C. Am. (5).
  Dasypodidae, armadillos, Am. (20).

ORDER INSECTIVORA   Insectivores (343)
  Solenodontidae, solenodons, C. Am. (2).
  Tenrecidae, tenrecs, otter-shrews, Mad., W. C. Af. (33).
  Chrysochloridae, golden moles, Af. (17).
  Erinaceidae, hedgehogs, moonrats, Eur., Af. (17).
  Soricidae, shrews, Eur., N. C. Am., Af. (245).
  Talpidae, moles, desmans, Eur., N. Am. (29).

ORDER SCANDENTIA   Tree shrews (16)
  Tupaiidae, tree shrews, S. E. As. (16).

ORDER DERMOPTERA   Flying lemurs (2)
  Cynocephalidae, flying lemurs, S. E. As. (2).

ORDER CHIROPTERA   Bats (950)
  Pteropodidae, Old World fruit bats, Af., S. As., Aus. (173).
  Rhinopomatidae, mouse-tailed bats, Af., S. As. (3).
  Emballonuridae, sheath-tailed bats, Wo. (50).
  Craseonycteridae, hog-nosed bat, Thailand (1).
  Nycteridae, slit-faced bats, Af., S. As. (11).
  Megadermatidae, false vampire bats, Af., S. E. As., Aus. (5).
  Rhinolophidae, horseshoe bats, O. W., Aus. (69).
  Hipposideridae, Old World leaf-nosed bats, Af., S. As., Aus. (61).
  Noctilionidae, bulldog bats, C. S. Am. (2).
  Mormoopidae, naked-backed bats, C. S. Am. (8).
  Phyllostomatidae, New World leaf-nosed bats, C. S. Am. (140).
  Desmodontidae, vampire bats, Am. (3).
  Natalidae, funnel-eared bats, C. S. Am. (8).
  Furipteridae, smoky bats, C. S. Am. (2).
  Thyropteridae, disk-winged bats, C. S. Am. (2).
  Myzopodidae, sucker-footed bat, Mad. (1).
  Vespertilionidae, vespertilionid bats, Wo. (319).

Table 1—*Continued*
   Mystacinidae, New Zealand short-tailed bat, New Z. (1).
   Molossidae, free-tailed bats, Wo. (91).

ORDER PRIMATES   Primates (179)
   Cheirogaleidae, dwarf lemurs, Mad. (7).
   Lemuridae, large lemurs, Mad. (16).
   Indriidae, leaping lemurs, Mad. (4).
   Daubentoniidae, aye-aye, Mad. (1).
   Lorisidae, lorises, galago, Af., S. E. As. (12).
   Tarsiidae, tarsiers, S. E. As. (3).
   Callithricidae, marmosets, tamarins, S. C. Am. (17).
   Cebidae, New World monkeys, S. C. Am. (32).
   Cercopithicidae, Old World monkeys, Af., S. E. As. (76).
   Pongidae, apes, W. C. Af., S. E. As. (10).
   Hominidae, man, Wo. (1).

ORDER CARNIVORA   Carnivores (240)
   Canidae, dogs, foxes, Eur., Am., Af. (35).
   Ursidae, bears, Eur., N. Am. (7).
   Procyonidae, raccoons, coati, kinkajou, olingo, Am. (18).
   Ailuropodidae, pandas, As. (2).
   Mustelidae, stoats, weasels, otters, badgers, skunks, Am. Af., Eur. (67).
   Viverridae, civets, genets, mongooses, Eur., Af., Mad. (72).
   Hyaenidae, hyaenas, Af., S. W. As. (4).
   Felidae, cats, Am., Eur., Af. (35).

ORDER PINNIPEDIA   Pinnipedes (34)
   Otariidae, sea lions, S. Hem., N. Pacific (14).
   Odobenidae, walrus, Arc. (1).
   Phocidae, seals, Wo. (19).

ORDER CETACEA   Whales, dolphins (79)
   Platanistidae, river dolphins, S. Am., S. E. As. (5).
   Delphinidae, dolphins, Wo. (32).
   Phocoenidae, porpoises, S. N. temperate, Arc. (6).
   Monodontidae, narwhal, white whale, Arc. (2).
   Physeteridae, sperm whales, Wo. (3).
   Ziphiidae, beaked whales, Wo. (18).
   Eschrichtidae, grey whale, N. Pac. (1).
   Balaenopteridae, rorquals, Wo. (6).
   Balaenidae, right whales, Wo. (3).

ORDER SIRENIA   Sea cows (4)
   Dugongidae, dugong, Indian Ocean coasts (1).
   Trichechidae, manatees, Atlantic coasts (3).

ORDER PROBOSCIDEA   Elephants (2)
   Elephantidae, elephants, Af., S. E. As. (2).

ORDER PERISSODACTYLA   Odd-toed ungulates (17)
   Equidae, horses, zebras, Af., W. C. As. (8).
   Tapiridae, tapirs, C. S. Am., S. E. As. (4).
   Rhinocerotidae, rhinoceroses, Af., S. E. As. (5).

ORDER HYRACOIDEA   Hyraxes (5)
   Procaviidae, hyraxes, Af., Arabia (5).

ORDER TUBULIDENTATA   Aardvark (1)
   Orycteropodidae, aardvark, Af. (1).

ORDER ARTIODACTYLA   Even-toed ungulates (184)
   Suidae, pigs, Eur., Af. (8).
   Tayassuidae, peccaries, Am. (3).
   Hippopotamidae, hippopotamuses, Af. (2).
   Camelidae, camels, llamas, S. Am., Af., As. (4).
   Tragulidae, mouse-deer, W. C. Af., S. E. As. (4).
   Moschidae, musk deer, E. As. (3).
   Cervidae, deer, Am., Eur. (34).
   Giraffidae, giraffe, okapi, Af. (2).
   Antilocapridae, pronghorn, N. C. Am. (1).
   Bovidae, cattle, antelopes, gazelles, sheep, goats, N. Am., Af., Eur. (123).

ORDER PHOLIDOTA   Pangolins, scaly anteaters (7)
   Manidae, pangolins, scaly anteaters, Af., S. As. (7).

ORDER RODENTIA   Rodents (1591)
   Aplodontidae, mountain beaver, N. Am. (1).
   Sciuridae, squirrels, marmots, Eur., Af., Am. (246).
   Geomyidae, pocket gophers, Am. (37).
   Heteromyidae, pocket mice, Am. (63).
   Castoridae, beavers, N. Am., Eur. (2).
   Anomaluridae, scaly-tailed squirrels, W. C. Af. (7).
   Pedetidae, spring hare, Af. (1).
   Muridae, rats, mice, voles, gerbils, Wo. (1011).
   Gliridae, dormice, Eur., Af. (14).
   Seleviniidae, desert dormouse, As. (1).
   Zapodidae, jumping mice, N. Am., N. Eur. (10).
   Dipodidae, jerboas, Af., As. (27).
   Hystricidae, Old World porcupines, Af. (12).
   Erithizontidae, New World porcupines, Am. (19).
   Caviidae, guinea pigs, S. Am. (15).
   Hydrochoeridae, capybara, S. C. Am. (1).
   Dinomyidae, pacarana, S. Am. (1).
   Dasyproctidae, agoutis, pacas, C. S. Am. (12).
   Chinchillidae, chinchillas, viscachas, S. Am. (6).
   Capromyidae, hutias, coypu, C. S. Am. (12).
   Octodontidae, degus, coruro, S. Am. (7).
   Ctenomyidae, tuco-tucos, S. Am. (32).
   Abrocomidae, chinchilla-rats, S. Am. (2).
   Echimyidae, American spiny rats, C. S. Am. (45).
   Thryonomyidae, cane rats, Af. (2).
   Petromyidae, African rock rat, Af. (1).
   Bathyergidae, African mole-rats, Af. (9).
   Ctenodactylidae, gundis, Af. (5).

ORDER LAGOMORPHA   Lagomorphs (54)
   Ochotonidae, pikas, N. C. As., N. Am. (14).
   Leporidae, rabbits, hares, Eur., Af., N. C. Am. (40).

ORDER MACROSCELIDEA   Elephant-shrews (15)
   Macroscelididae, elephant-shrews, Af. (15).

CHAPTER TWO

GEOGRAPHICAL DISTRIBUTION

The mammals are a widespread group of animals. Terrestrial species extend from the Arctic to the southern tip of South America and many islands of the southern oceans. They ascend mountains up to and into the snow line, and inhabit the driest and most arid deserts. The marine species exploit all the seas, from the cold north to the cold south. There is no continent on which mammals are not found. However, despite this broad geographical spread, there are discernible patterns in the present distribution. For example, it is common knowledge that kangaroos are found only in Australia, giraffes in Africa and giant pandas in China. It is also implicit from these observations that ranges are limited and that impediments must exist, preventing the free movement of species from one place to another.

In the following account the major zoogeographical regions will be identified, as well as the faunas they support. Similarities and differences between regions will be highlighted and in the process the extent of endemism, limited distribution and ubiquity will become apparent. The geographical range of a species can be static or it may change, and it is appropriate to examine the broad ecological and physical circumstances that facilitate or prevent alteration of range. The present distribution of mammals is principally the result of changes in the structure of the earth's surface during the Tertiary, previous climate and vegetation, as well as the evolution throughout this time of the mammals themselves. This takes us into the fascinating study of historical zoogeography which is beyond the scope of our present interests.

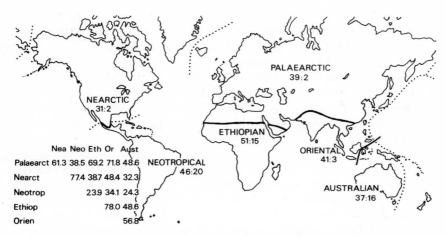

Figure 2.1 World zoogeographic regions and indices of similarity at familial level between regions. Index of similarity = $100C/N$, where C is the number of families common to both regions and N is the number of families in the smaller fauna. The total and number of endemic families are indicated for each region.

2.1 Zoogeographical regions

In 1876 Wallace subdivided the world's land masses into six geographical regions (Fig. 2.1). The division was based on the precept that the fauna within a region had a distinctive identity that separated it from the fauna of other regions. Although our knowledge of mammal distributions has greatly increased over the past century (Corbet and Hill, 1980), these regions still remain useful entities for zoogeographical study of the group. Even so it is important to recognize that they have limitations and inevitably the drawing of lines on maps does not necessarily have biological reality.

Palaearctic Region. This is the largest of the regions, including all Europe, much of Asia and North Africa, and is mainly temperate although its extension into the Arctic in the north and the subtropics of the Sahara Desert in the south ensure a wide range of conditions. The Palaearctic is separated from the Oriental Region by the mountainous massifs to the north of the Indian subcontinent. It has a moderately rich fauna with 39 families of which two, Seleviniidae (desert dormice) and Ailuropodidae (pandas) are endemic. Orders such as the Chiroptera, Insectivora, Rodentia, Carnivora and Artiodactyla are well represented.

Nearctic Region. This region encompasses North America and like the preceding region is mainly temperate. Again extensions north to the Arctic and south into sub-tropical Mexico add to the variety of habitats. Two families, the Aplodontidae (mountain beavers) and Antilocapridae (pronghorn antelopes) are endemic. Primates (except man) are absent, but there is considerable variety of bats and rodents and a good representation of carnivores, insectivores and artiodactyls. Among the groups of limited world distribution are the opossums (Didelphidae) and armadillos (Dasypodidae).

Neotropical Region. This extends from South and Central America to Mexico where it meets the Nearctic Region. Tropical grassland and forest cover much of the land mass, which, together with the southern extension into cold climates and alpine tundra and mountain forests of the high Andean mountains, results in appreciable richness of habitat. The region has the second largest number of families and as twenty of them are endemic there is more endemism than in any other region. Among these are two families each of marsupials (colocolos, shrew opossums), Edentata (anteaters, sloths), primates (marmosets, New World monkeys) and ten families of rodents. The last include chinchillas, capybaras, agoutis, toucotoucos and guinea pigs. The Artiodactyla are not well represented although there are the unique llamas, vicunas and peccaries.

Ethiopian Region. This region extends from south of the Atlas Mountains to the southern tip of Africa. It includes part of the Arabian Peninsula and the island of Madagascar. Within these boundaries are hot deserts, tropical forests and savannas and high mountains with their complex variety of vegetation. With 51 families this region contains more than any other. Among the 15 endemic families are the golden moles (Chrysochloridae), tenrecs (Tenrecidae), lemurs (Lemuridae, Indriidae), aardvarks (Orycteropidae), giraffes (Giraffidae), six families of rodents but only one of bats. Perhaps most characteristic of this region is the multiplicity of species of ungulates which inhabit the savannas. The fauna of Madagascar is remarkably dissimilar from the rest of the region. Of 57 genera and 102 species, 40 and 83 respectively are endemic. Two species, a civet (*Viverricula*) and the bush pig (*Potamochoerus*) were probably introduced by man. The 12 genera of primates are represented by the lemurs and the aye-aye (*Daubentonia*); the tenrecs comprise all but two species of the Insectivora, the Carnivora are represented by a unique assemblage of viverrids and the rodents by an endemic subfamily of Muridae. Only the bats share more than half their species with the mainland. Not only is the Madagascar fauna unique; it has a very different composition from that of the mainland. This is evident from the absence of ungulates, cats, canids, rodents such as porcupines and dormice, spring hares and monkeys.

Oriental Region. This region includes Asia south of the Himalayas and a line extending south-east from them to the shores of the China Sea. Among the islands included are those east and north of a line (Wallace's Line) between Borneo and Sulawesi and Bali and Lombok. The region includes extensive areas of tropical lowland rain forest. There are only three endemic families. They are the flying lemurs (Cynocephalidae), musk deer (Moschidae) and the hog-nosed bats (Craseonycteridae). There are resemblances with other regions. For example the Lorisidae, Pongidae, Manidae, Elephantidae and Rhinocerotidae occur here and in the Ethiopian region; but as the oriental region is predominantly forested there is not the variety of ungulates found in Africa. The drawing of an arbitrary line through an archipelago has inevitably produced anomalies.

Australian Region. This includes Australia and its offshore islands as well as those of the East Indies up to Wallace's Line. The land mass of Australia provides considerable topographic variation, the climate is tropical to sub-tropical and the annual precipitation variable from one part of the continent to another. The range of available habitats is exploited by monotremes and marsupials and all those present are endemic to this region. Placentals are also present. Apart from recent introductions by man there are several genera of bats and rodents on the main island. The islands linking Australia and the Oriental mainland show a gradual transition in their faunal composition (section 2.4). The inclusion of some of these islands in the Australian region inflates the number of widespread families of placentals present and in so doing masks the considerable endemism within the continental land mass.

Measurements of similarity (Simpson's $100C/N$, see Fig. 2.1) at family level, between the regions (Fig. 2.1) illustrates the extent to which each region has families in common with other regions. No region has a unique fauna. Some families are ubiquitous—these include the Emballonuridae, Vespertilionidae and Molossidae among the bats, the Hominidae among the Primates and the Muridae among rodents. But many of the similarities result when a family is shared between a limited number of regions. It is also evident that some regions have close affinities not shared by other regions, for instance the Oriental and Ethiopian, the Palaearctic and

Oriental, the Nearctic and Neotropical, and the Ethiopian and Palaearctic. These measurements provide a guide to contemporary distributions but are in themselves inadequate to explain the causes of the configuration.

2.2 Ecological equivalents

The successful exploitation by the mammals of a wide range of habitats and microhabitats is due to the appearance of ecological specialists in each of the major land masses—such forms as ant-eaters, cursorial (running) herbivores, small and large flesh-eaters and fruit-eaters. To achieve this, the same evolutionary processes have not always been applied. In some instances members of the same family with similar structural adaptations and ecological requirements have spread to more than one region. This is illustrated by the pigs, which typically feed on ground and subsurface vegetation which they root out with the forelegs and snout. There are the peccaries of North America, the bush pigs and warthogs of Africa and the wild boars of Europe and Asia. In other instances equivalent ecological roles may be filled by mammals of appreciably less close phylogenetic affinity. The anteaters of Africa are represented by the aardvark (order Tubulidentata) and pangolin (order Pholidota), in South America by the Myrmecophagidae (order Edentata), in Australia by the numbat (order Marsupialia) and in south-east Asia by the pangolin. The mammals filling this niche are from four different orders but nevertheless have several

Table 2.1 Percentage of species occupying major adaptive zones in African and Neotropica. (After Keast, 1969).

	Africa	Neotropica
Small insect-eaters and predators (shrews, etc.)	9.1	2.2
Fossorial moles	1.7	—
Rodents in mole niche	—	0.1
Specialized ant and termite feeders	0.8	0.4
Small to medium-sized terrestrial omnivores	3.7	3.8
Typical murid rodents	21.4	27.5
Miscellaneous rodents, non-murid	2.8	12.6
Rabbit-sized herbivores	4.0	2.1
Medium to large terrestrial herbivores	12.6	2.8
Fossorial herbivores (underground feeding)	2.0	0.02
Arboreal herbivores, omnivores and insect-eaters	11.4	13.1
Carnivores, weasel to fox-size	4.8	4.0
Carnivores and scavengers, large	0.9	0.2
Bats, fruit and blossom-feeding	3.4	9.9
Bats, insectivorous	19.7	17.7

structural adaptations in common. Other examples of ecological convergence include the tree-dwelling leaf-eating sloths of South America (order Edentata) and the koala bear of Australia (order Marsupialia), and the moles (Talpidae) of the Palaearctic and Nearctic, the golden moles of Africa (Chrysochloridae) and the marsupial moles of Australia (Notoryctidae). The first two of the groups of moles have close affinity, belonging to a common order (Insectivora), unlike the third group.

Ecological interest centres particularly on the range of adaptations the mammals collectively display within a region or locality, and whether the faunas of each region cover a similar adaptive spectrum. Comparison of the percentage of species in different adaptive categories in Africa and Neotropica (Table 2.1) illustrates considerable similarity particularly when allowance is made for differences in habitat and proportions of the main vegetation types. Africa has relatively more large herbivores but this is balanced by the richness of the small herbivores (non-murid rodents) in South America. The insectivorous bats, arboreal herbivores and murid rodents occur in similar proportions and comprise over 50 % of the species present.

2.3 Mammals of the oceans

The oceans of the world cover 71 % of the earth's surface and, unlike the land areas, provide a continuous, uninterrupted habitat with free interconnection between one ocean and the next. Within this expanse are considerable physical and biological variations in environmental conditions which has led to the adaptation and specialization of the two groups of mammalian inhabitants. These are the seals, including sea-lions and walrus (order Pinnipedia) and the whales and their allies (order Cetacea). An important ecological difference between these two groups is that the Pinnipedia must come ashore to breed.

Each species of seal has a well-defined shoreline on which it produces its young and rears them through early life (Fig. 2.2). These coasts and their offshore waters describe the range of the species. Seals are cold-water animals—the 20°C summer isotherm in both hemispheres forms the distributional limit for most species. Exceptions are the West Indian, Mediterranean and Hawaiian monk seals (*Monachus*) which prefer slightly warmer waters (King, 1964).

The whales are more widespread, more mobile and include a greater range of size than the seals. The larger whales of the southern hemisphere—i.e. the blue fin (*Balaenoptera* spp.) and humpback

Figure 2.2   Distribution of southern fur seals of the genus *Arctocephalus* (from King, 1964).

(*Megaptera*)—each have favoured locations between 35°S and 65°S in the summer months. In winter they migrate north to warmer waters. The northern hemisphere populations of these species make similar movements between warmer and colder regions. Thus both populations occur in the milder equatorial waters, but, because of climatic differences in the two hemispheres, not at the same time. These and several other species including sei (Balaenoptera), sperm (*Physeter*), bottlenose (*Hyperoodon*), killer (*Orcinus*) and pygmy sperm whales (*Kogia*) are migrating, oceanic cosmopolitans. In contrast, several cetaceans have much more restricted distributions. Some (e.g. *Stenella, Sotalia, Steno*) are found in many tropical and sub-tropical seas; others are even more restricted, such as the Irawadi dolphin (*Orcaella*) found only in the Bay of Bengal and off Malacca and Thailand. Finally, some species are restricted to the cold north, e.g. Beluga (*Delphinapterus*) and narwhal (*Monodon*), and others to the cold south, e.g. Commerson's dolphin (*Cephalorynchus*) (Slijper, 1962).

2.4 Barriers to distribution

Consideration can now be given to the situations and circumstances which restrict and permit the exchange of faunas between two locations. These paths of faunal interchange fall into three categories (Simpson, 1965). Firstly, there is the situation where impediments, both physical and ecological, are few, so that gradual dispersal could proceed with little restriction. Interchange along such a route would be unselective and would permit movement in both directions from the two sources. A large area permitting so free a movement is the land mass of Eurasia. Secondly, the situation where limited interchange is possible, where there is a partial barrier or filter, is found where two land masses are connected, possibly by

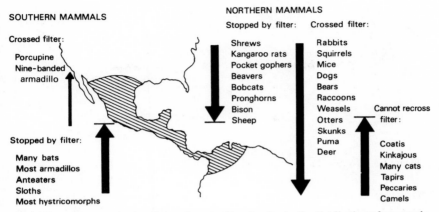

Figure 2.3 The role of Central America as a filter barrier to the distribution of mammals from north and south (modified from Simpson, 1965).

a narrow neck of land covered by a particular vegetation type. The species moving from one area to another can do so only if they are adapted to this vegetation. The isthmus between North and South America, which at its narrowest is covered by rain forest, illustrates this, and has resulted in selective infiltration (Fig. 2.3) in both directions. The present composition of the faunas of North and South America has, however, been influenced not only by the contemporary connection but also by earlier Tertiary connections and disconnections when climatic and vegetational conditions may not have been the same as at present. The third path of faunal interchange is the "sweepstake route". Here distributional barriers are considerable. In a sweepstake there is invariably a winner but it is a matter of chance who that winner is. In the distribution of animals this is found in archipelagos, where barriers are considerable, but nevertheless some species make the hop, by some means, from mainland to island or from island to island. This is well illustrated by the faunas of the East Indies (Fig. 2.4). The marsupials originating from Australia decline in number of genera going westwards and the placentals, excluding rodents and bats, going eastwards. The situation is complicated because the islands have different land areas and the larger islands are capable of supporting a more diverse fauna. Much of the placental fauna has its origin in south-east Asia. The fauna of Borneo is typically Asian, having no marsupials and many more genera of placentals than Sulawesi.

Examination of these insular faunas highlights the difficulty in drawing meaningful lines through archipelagos to separate zoogeographic regions.

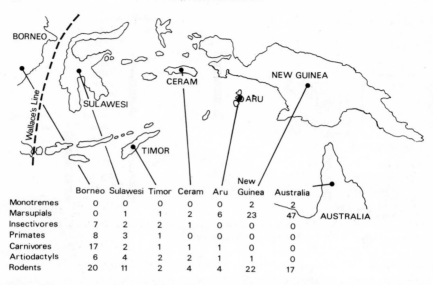

	Borneo	Sulawesi	Timor	Ceram	Aru	New Guinea	Australia
Monotremes	0	0	0	0	0	2	2
Marsupials	0	1	1	2	6	23	47
Insectivores	7	2	2	1	0	0	0
Primates	8	3	1	0	0	0	0
Carnivores	17	2	1	1	1	0	0
Artiodactyls	6	4	2	2	1	1	0
Rodents	20	11	2	4	4	22	17

Figure 2.4 Number of genera of terrestrial mammals in Australia and islands of the East Indies (from Laurie and Hill, 1954; Corbet and Hill, 1980).

All the islands except Borneo in Fig. 2.4 are within the Australasian region as Wallace's Line passes west of Sulawesi. In fact, Sulawesi has closer affinity with Asia, even though it has a genus of marsupial (*Phalanger*). The presence of the tarsiers on both sides of Wallace's Line also suggests an unnatural division.

Emphasis has so far been placed on distributional limitations between extensive geographical areas. More detailed consideration can now be given to the operation of ecological barriers within a continental land mass. The African rain forest extends as a continuous belt from the west coast to the Rwenzori mountains. Further east, isolated patches of forest are found where climate and edaphic conditions are suitable (Fig. 2.5). Some of these forests cover large areas and provide a good high canopy although their floral composition often differs from the western forests. Nevertheless, they afford a suitable habitat for forest species. In the past there may have been forested connections between these eastern blocks and the main forest, so that the contemporary distribution should not be accounted for on the basis of hops made between them. It is true, though, that the major recent barriers to distribution have been the intervening grasslands and savannas and these have been effective in preventing an

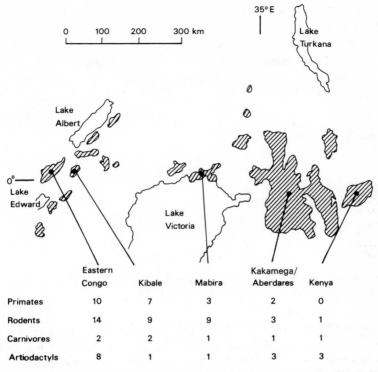

	Eastern Congo	Kibale	Mabira	Kakamega/ Aberdares	Kenya
Primates	10	7	3	2	0
Rodents	14	9	9	3	1
Carnivores	2	2	1	1	1
Artiodactyls	8	1	1	3	3

Figure 2.5 The number of exclusively forest species in various forests in East Africa. These faunas have probably originated from the Eastern Congo forest with a progressive eastward decline in species numbers (from Misonne, 1963).

easy movement of forest mammals (Fig. 2.5). If interchange had been readily possible, then these forest faunas should resemble each other appreciably more closely than they do.

2.5 Species abundance

The number of species in an area is frequently a reflection of the diversity of terrain and the richness of vegetation it supports. Comparison of selected faunas of Europe with those of tropical Asia and Africa (Table 2.2) suggests greater species richness in the tropics. In all localities bats, rodents and carnivores feature prominently. The faunistic richness of Zaïre must be accounted for by the wide range of altitudes (up to 5100 m) and climatic

Table 2.2 Number of species of terrestrial mammals (after Delany and Happold, 1979)

Country	Tropical Africa		Tropical Asia	Temperate Europe	
	Zaïre	Zambia	Malaya	Europe	France and Switzerland
Area (ha × 1000)	235 690	74 498	13 126	c. 453 000	59 322
Insectivora[1]	49	16	7	17	14
Dermoptera	—	—	1	—	—
Chiroptera	116	55	81	31	26
Primates	32	7	13	1[2]	—
Carnivora	36	28	30	24	14
Pholidota	3	2	1	—	—
Lagomorpha	2	4	—	3	3
Rodentia	95	63	54	48[3]	23[3]
Tubulidentata	1	1	—	—	—
Hyracoidea	5	2	—	—	—
Proboscidea	1	1	1	—	—
Perissodactyla	3	2	3	1	1
Artiodactyla	42	28	9	13[4]	7[4]
Total	385	209	200	138	88

[1] Includes Macroscelidea.
[2] The Barbary Ape (*Macaca sylvanus*), probably an introduction.
[3] Excludes the introduced grey squirrel (*Sciurus carolinesis*) and coypu (*Myocastor coypus*).
[4] Does not include any of the following deer that have been introduced into Europe: Chinese water deer (*Hydropotes inermis*), muntjac (*Muntiacus reevesi*), sika deer (*Sika nippon*) and white-tailed deer (*Odocoileus virginianus*).

conditions, providing a considerable range of forest and grassland vegetation; Zambia is largely covered by wooded savanna and Malaya by lowland and montane forest. Temperate Europe in spite of its extensive land coverage and variety of vegetation and terrain supports a modest fauna. However, in France and Switzerland (which amount to only 13 % of the land area examined) 64 % of the fauna is represented.

Turning to North and Central America, there is a general increase in species numbers from north to south (Fig. 2.6) with 30 to 40 species in the Arctic and over 140 in the tropics. There are further features of note. Mountainous areas, whether in the Appalachians or Rockies, support larger numbers of species. There is also a trend to higher speciation in the west independent of topography. There are a number of fronts of abrupt change, frequently (but not invariably) associated with mountainous areas. Finally, species numbers are low on peninsulas. There is thus general

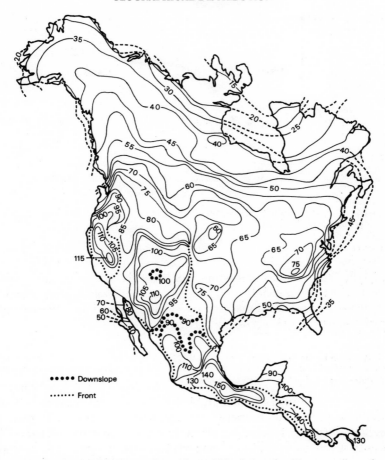

Figure 2.6 Species density contours in continental North America. Fronts are lines of rapid change (modified from Simpson, 1964).

agreement that warmer climates sustain more species than their temperate equivalents.

The relationship between altitude and species abundance has been examined in the rodents of western Rwenzori. At the base of these mountains are areas of lowland forest and savanna. As they are ascended there is a succession of mountain forest, bamboo forest, ericaceous scrub and finally afro-alpine vegetation up to the peak of over 5000 m. The number of rodents declines from 35 species at the base (700 m) to 3 at

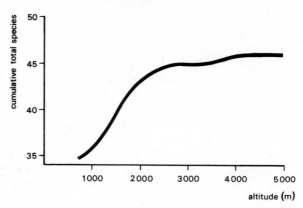

Figure 2.7   Cumulative increase in number of rodent species in western Rwenzori, Zaïre with increasing altitude (from Delany, 1972).

4400 m and none at 4700 m (Fig. 2.7). There is a total of 46 species on the mountain and in its foothills with 11 of these species being added above 1000 m. Thus the appearance of the Rwenzori massifs has resulted in a 24 % increase in species abundance above the lowland level with the major addition of new species occurring between 1000 and 2500 m.

2.6 Distributional changes

The range of a species can alter. This may be due to changes in the habitat which produce more or less favourable conditions, or to a species crossing previously restrictive barriers and entering a favourable habitat. Considerable range contraction of mammals can be attributed to man's activities. Local extinction has been and still is widespread. The wolf and wild boar disappeared from Britain by the mid-eighteenth century, the North American bison experienced great loss of range in the nineteenth century, while human harassment is now causing appreciable range contraction of many large African mammals.

   Introductions, both accidental and deliberate, by man have greatly increased the range of some species. Perhaps foremost among these are the spread of house rats (*Rattus*) and mice (*Mus*) through much of the inhabited world. Introductions for amenity and commerce have resulted in numerous establishments of exotics. In Britain, six species of deer have arrived in this way, as well as the grey squirrel, fat dormouse, red-necked wallaby, rabbit and American mink. The most striking example of a mammal fauna comprised almost exclusively of introduced species is that of New Zealand. Unlike that of Britain, the New Zealand native fauna is very small, consisting of two species of bat. The introductions, mainly from Europe and Australia, number 33 species of wild and feral mammals including an opossum, six wallabies, a European hedgehog, a rabbit and a hare, four rodents, four carnivores and eight deer. Several, such as rodents, rabbits and red deer, have attained pest status (Gibb and Flux, 1978).

Natural range extensions can occur and have been observed in the red fox (*Vulpes*) (MacPherson, 1964). This entered Baffin Island in the Canadian Northwest Territories about 1918, and spread through the island and on to more northerly islands, which it presumably reached across the winter ice. From information on years of first appearance in this new range its rate of spread has been estimated at 21.5 km per year. A second example is the nine-banded armadillo (*Dasypus novemcinctus*) (Humphrey, 1974). This animal has increased its range within the United States since the turn of the century when it had a localized distribution in southern Texas. Since then it has occupied a broad arc of the central and southern states from Oklahoma to Florida.

CHAPTER THREE

REPRODUCTION AND LIFE HISTORIES

An understanding of the reproductive biology of a species allows the prediction and estimation of the number of young it will produce, while a knowledge of the life history provides information on the age at which animals reproduce and the length of time during which they are likely to contribute new animals to the population. Each species of mammal has physiological characteristics that determine its pattern of reproduction but, as will be seen, these can be modified within limits in response to environmental stimuli such as climate and food supply. Similarly, during a life history, which is the sequence of events through which an animal passes from fertilized egg to death, development stages can be abbreviated or extended according to ecological conditions. While gross reproductive and life history characteristics are highly relevant, detailed physiological processes are beyond the requirements of the present account.

3.1 The sexual cycle

The adult female normally sheds eggs (*ovulation*) from the ovary at regular intervals. The number varies both within and between species. The former can result from age differences between females as well as variation in habitat conditions. Ovulation may take place spontaneously (*spontaneous ovulation*) or in response to the stimulus of copulation (*induced ovulation*). The latter is infrequent, being found in some carnivores, rodents and lagomorphs. Ovulation is accompanied by a thickening and vascularization of the uterus wall in anticipation of receiving a fertilized egg and the production of an embryo. If this does not happen, the uterus resumes its normal shape and, after an interval, the ovary sheds a further batch of eggs. In primates there is a breakdown of cells as the uterus returns to its non-receptive condition which results in external bleeding. The intervals between ovulation are regular. The cycle of changes within the reproductive system is the *oestrous cycle*, except in primates, where it is a *menstrual cycle*. The period of receptivity is the oestrus. Some mammals have only one oestrous cycle (*monoestrous*) in the breeding season while others have several (*polyoestrous*). Reproductive capability is increased in some species which have an oestrus within a few hours of giving birth (*post-partum oestrus*). This is common in rodents and lagomorphs and the result is that the female is simultaneously suckling and pregnant. This could put a considerable strain on maternal resources, particularly if lactation is of the same duration as gestation. The African tree rat (*Thamnomys*) adapts to this situation by not mating at the post-partum oestrus but waiting until the next oestrus a few days later (Bland, 1973). Similarly, in the African elephant a minimum 30-month calving interval occurs after a 22-month gestation period (Laws, 1968a). In most placentals the oestrous cycle is suppressed during gestation but in the majority of marsupials as well as some placentals, e.g. the fox, the period of gestation is shorter than the length of the oestrous cycle.

Sexual activity in the male commences with the maturation of the testes and the production of spermatozoa. As with the female, the attainment of maturity can be determined by a range of environmental factors. Both sexes may have periods of sexual inactivity or quiescence after the attainment of maturity.

The potential production of young is extremely variable (Table 3.1). In general, small mammals, with their early maturation, short gestation and lactation periods, frequent oestrus and large litters, have considerable potential for increase. The majority of marsupials do not have shorter gestation periods than eutherians of equivalent size—the opossum, *Didelphis marsupialis*, and grey kangaroo, *Macropus giganteus*, are exceptional (Tyndale-Biscoe, 1973). Many cetaceans, including the enormous blue whale (*Balaenoptera musculus*) have relatively short gestation periods for their size, and as they produce 1 or 2 calves every two years, have a potential comparing favourably with the terrestrial elephant and hippopotamus.

3.2 Reproductive patterns

There are three variants of the reproductive pattern just described which extend the interval between the production of sexual products by at least one of the sexes, and parturition. These are delayed fertilization, delayed

Table 3.1 Reproductive data of selected mammals

Species	Age at first conception	Polyoestrous (P) or monoestrous (M)	Frequency of oestrus or menstruation (days)	Litter size	Gestation period (days)	Lactation
Marsupialia						
Long-nosed bandicoot (*Perameles nasuta*)	265 dy.	P	13–28	1–4	12	55–62 dy.
Mouse opossum (*Marmosa robinsoni*)	17 mo.	P	18–31	7–9	14	60–70 dy.
Euro (*Macropus robustus*)		P	35	1	33	235 dy.
Insectivora						
European hedgehog (*Erinaceus europaeus*)	11 mo.	P		2–7	31–35	4–6 wk.
Common shrew (*Sorex araneus*)	8–9 mo.	P		5–7	13–19	19 dy.
Chiroptera						
Fruit bat (*Eidolon helvum*)	?1 yr.	M		1	120	30 dy.
Primates						
Bushbaby (*Galago senegalensis*)	1 yr.	P	32	1–2	124–126	?3.5 mo.
Chimpanzee (*Pan troglodytes*)	7–10 yr.	P	37	1–2	227	2–4 yr.
Cetacea						
Blue whale (*Balaenoptera musculus*)	8–10 yr.	P?		1–2	360	6–7 mo.
Common porpoise (*Phocaena phocaena*)	3–4 yr.	P?		1	335	8 mo.
Proboscidea						
African elephant (*Loxodonta africana*)	11–20 yr.	P	14–21	1	670	2 + yr.
Carnivora						
Red fox (*Vulpes vulpes*)	10 mo.	M		1–10	52–53	6 wk.
American mink (*Mustela vison*)	10–11 mo.	P	7–10	3–6	45–52	8 wk.
Lion (*Panthera leo*)	3–4 yr.	P	11–65	1–6	105–113	8 mo.
Rodentia						
Deer mouse (*Peromyscus maniculatus*)	c. 6 wk.	P	4–5	2–9	22–27	30–37 dy.
House mouse (*Mus musculus*)	c. 6 wk.	P	4–6	4–8	19–20	18–20 dy.
Artiodactyla						
Hippopotamus (*Hippopotamus amphibius*)	9 yr.	P	20	1–2	340	12 wk.
Red deer (*Cervus elaphus*)	3 yr.	P	18	1–2	227–236	8–10 mo.
Eland (*Taurotragus oryx*)	28 mo.	P	21	1–2	225–280	

implantation and delayed development. As will be seen, their adaptive significance is frequently apparent, although in some instances the reasons for adopting these strategies remain far from clear.

Delayed fertilization is found in some bats from north temperate regions. They usually hibernate, and during this time their sexual organs are quiescent and body reserves decline. In order to maximize use of resources in spring and summer they need to start breeding as soon as possible after hibernation. But this is also a time when they have to build up their body tissues and as a result breeding could be delayed. This problem has been circumvented by the production of spermatozoa in the autumn, after which the testes regress and sperm is stored in the caudal epididymides. This is viable for up to seven months. As the sperm can be retained in the female tract, copulation takes place in autumn with immature and non-reproductive adult females. When these females become sexually active in the following spring fertilization is immediate. The phenomenon has been found in New and Old World genera including *Myotis*, *Pipistrellus*, *Eptesicus* and *Antrozous* (Wimsatt, 1945; Racey, 1973). The advantages of delayed fertilization are not as apparent in non-hibernating tropical vespertilionids which also store sperm (Yalden and Morris, 1975).

In delayed implantation the fertilized egg divides to form a blastocyst which then delays further development until its implantation into the wall of the uterus. It has been recorded in several orders including the Carnivora (Fig. 3.1) where it is particularly common amongst mustelids, Chiroptera, Rodentia, Marsupialia, Edentata and Artiodactyla. In some species, e.g. fruit bat (*Eidolon*) (Mutere, 1967), polar bear (*Thalarctos*) (Volf, 1963), and roe deer (*Capreolus*) (Short and Hey, 1966), the length of the delay is constant. However, in the badger (*Meles*) in Britain (Neal and Harrison, 1958), and spotted skunk (*Spilogale*) (Mead, 1968a, b), it is variable. Female badgers experiencing a shorter delay are those that have recently matured and older animals where earlier matings did not result in fertilization. Delays in the skunk are shorter in the eastern United States than in the west.

Australian marsupials delay implantation as long as they have a developing animal in the pouch. The quokka (*Setonix brachyurus*) mates in January and after a 28 day gestation the young leaves the uterus and enters the pouch. There is then a post-partum oestrus which frequently results in fertilization. The pouch animal is normally suckled until August and during this time the embryo develops to the blastocyst stage. Further development is not continued until the young animal leaves the pouch

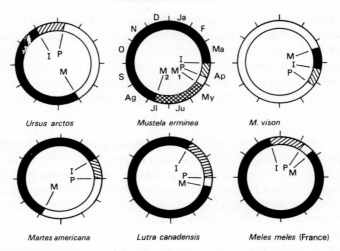

Figure 3.1 Delayed implantation cycles in brown bear (*Ursus arctos*), stoat (*Mustela erminea*), American mink (*Mustela vison*), American marten (*Martes americana*), Canadian river otter (*Lutra canadensis*) and European badger (*Meles meles*). Months of the year are shown in the upper central figure. M, I and P show dates of mating, implantation and parturition. Period of delay in black; gestation stippled (modified from Ewer, 1973).

Table 3.2　Gestation periods and litter sizes in carnivores

Species	Gestation period (days)	Litter size Range	Usual or mean (m)	Notes
Canidae				
Wolf (*Canis lupus*)	63	3–11	6.5 (m)	
Hunting dog (*Lycaon pictus*)	72–73	2–12	7	
Red fox (*Vulpes vulpes*)	51–52	1–9	5.4 (m)	New York
Mustelidae				
Stoat (*Mustela erminea*)	21–28[1]	4–10	6.4 (m)	USA
Weasel (*Mustela nivalis*)	43–47	4–8	6.2 (m)	UK
Pine marten (*Martes martes*)	28[1]	2–5	3	
Badger (*Meles meles*)	42[1]	1–4	2, 3	Europe
Sea otter (*Enhydra lutris*)	?	1–2	1	
Striped skink (*Mephitis mephitis*)	62–66	5–9	6, 7	
Hyaenidae				
Spotted hyaena (*Crocuta crocuta*)	110	1–3	2	
Felidae				
Cheetah (*Acinonyx jubatus*)	90–95	3–4	3	
Lion (*Panthera leo*)	105–113	1–6	3, 4	

[1]Excluding period as unemplanted blastocyst

whether this be prematurely or when it has completed its full term. The resting stage is referred to as a diapause (Sharman and Berger, 1969). In practice few animals rear a second offspring in August as by this time most of the blastocysts have died and the female is in anoestrus. The main benefit of this system is the provision of a replacement animal for those lost from the pouch earlier in the season (Shield and Woolley, 1963). While the red kangaroo (*Megaleia rufa*) also has a diapausing blastocyst, it differs from the quokka in breeding throughout the year and having a high survival of the blastocyst which commences development 30 days before the termination of 230 days pouch life of the young (Sharman, 1970).

Delayed development is a little-documented phenomenon recorded in bats. Following fertilization the embryo develops and the blastocyst implants in the uterine wall. There is then a delayed development. This has been recorded in the Jamaican fruit bat *Artibeus* (Fleming, 1971), where the young appear when fruit availability is at a maximum and in the European vespertilionid, *Miniopterus schreibersi*. It may also occur in the North American phyllostomatid *Macrotus* (Yalden and Morris, 1975).

3.3 Litter size

An examination of litter size in the main orders indicates that some, e.g. primates, Chiroptera, Cetacea, Pinnipedia, Proboscidea, Perissodactyla and Artiodactyla, usually produce small litters, frequently after relatively long gestation periods. Here the production of a single young animal is typical and more than two is uncommon. Exceptions are relatively few, and include some vespertilionid bats, pigs and deer which can produce up to four, fourteen and six young respectively. Within each of the remaining major orders (Marsupialia, Carnivora, Rodentia) there is a remarkably wide range of litter size. Marsupials such as the kangaroos and wallabies (Macropodidae) and the Australian opossums and cuscuses (Phalangeridae) have small litters of one to three young. In contrast, the American opossums (Didelphidae) produce litters of from 10 to 25 individuals. In the Carnivora (Table 3.2) litters range from typically one in the sea otter (*Enhydra*) to seven in the wild dog (*Lycaon*). Among the rodents (Table 3.3) both large, e.g. *Atherurus*, and small, e.g. *Otomys* species, may have small litters, while large rodents such as coypus (*Myocaster*) and capybara (*Hydrochoerus*) may have larger litters of similar size to those in many species of the smaller rats and mice.

There are numerous examples of intraspecific variation in litter size. These can be related to intrinsic factors such as maternal age and extrinsic

Table 3.3 Gestation periods and litter sizes in rodents

Species	Gestation period (days)	Litter size Range	Usual or mean (m)
Sciuridae			
Grey squirrel (*Sciurus carolinensis*)	42–45	1–8	3
Geomyidae			
Pocket gopher (*Geomys bursarius*)	40–50	1–9	4
Heteromyidae			
Merriam's pocket mouse (*Dipodomys merriami*)	33	2–4	3
Muridae			
Grooved-toothed rat (*Otomys denti*)		1–2	1
Bank vole (*Clethrionomys glareolus*)	17–18	3–5	4.1 m
Nile rat (*Arvicanthis niloticus*)	18	3–10	4
House mouse (*Mus musculus*)	19–20	4–8	5, 6
Brown rat (*Rattus norvegicus*)	21–24	4–10	7, 8
Multimammate mouse (*Praomys natalensis*)	23	1–19	12
Hystricidae			
Crested porcupine (*Hystrix cristata*)	42–64	1–4	2
Brush-tailed porcupine (*Atherurus africanus*)	100–110	1–3	1
Hydrochoeridae			
Capybara (*Hydrochoerus hydrochaeris*)	119–126	2–8	
Capromyidae			
Coypu (*Myocastor coypus*)	127–138	1–13	5.3 m
Bathyergidae			
Orange-toothed mole-rat (*Tachyoryctes splendens*)	37–40	1–4	1

factors such as geographical location, population density and other environmental conditions. There is considerable information on seasonal differences in litter size particularly among mammals having a short gestation period. In the Russian tundra, the vole *Microtus gregalis* has mean numbers of embryos per female of 6.6, 8.7 and 9.9 in winter, spring and summer generations respectively (Svarc *et al.*, 1969). The mean number of embryos in the wood mouse (*Apodemus*) in Britain are 5.5 in summer, 6.0 in mid-summer and 5.0 in late summer (Baker, 1930). A similar pattern has been recorded in the bank vole (*Clethrionomys*) which has the smallest litters of all in winter breeding females (Smyth, 1966). Litter size can also vary from year to year—rabbits (*Oryctolagus*) on the Island of Skokolm had mean annual litter sizes ranging from 3.4 to 4.8 in the years 1959 to 1963 (Lloyd, 1970).

Species with extensive geographical ranges may have different litter sizes in different parts of the range. The European mole (*Talpa*) most commonly

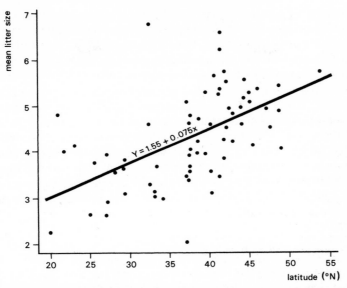

Figure 3.2 Regression of mean litter size and latitude in *Peromyscus* species in North America (modified from Smith and McGinnis, 1968).

has embryo numbers of four in Britain and Germany, five in Italy and six in the Ukraine (Godfrey and Crowcroft, 1960). The widespread multi-mammate mouse (*Praomys natalensis*) of tropical and sub-tropical Africa has mean embryo numbers of 9.5 in the Transvaal and 12.6 in Uganda (Delany, 1972). A more readily interpretable pattern of variation in litter size has been found in *Peromyscus* in North America where in 14 species there was a steady decline in numbers from high to low latitudes (Fig. 3.2). In both rodents and lagomorphs changes in litter sizes have been recorded with increasing altitude. In Colorado, there is an increase in embryo numbers of *Peromyscus maniculatus* from a mean of 4.0 at 1550 m to 5.6 at 3350 m (Dunmire, 1960). Embryo numbers of hares (*Lepus capensis*) in Kenya decline with increasing altitude from 1.75 at 600 m to 1.24 at 1800 m (Flux, 1969).

Litter size can increase with maternal age. This has been recorded in mule deer (*Odocoileus*) (McConnell and Dalke, 1960), grey foxes (*Urocyon*) (Wood, 1958), cottontails (*Sylvilagus*) (Lord, 1961) and bank voles (*Clethrionomys*) (Brambell and Rowlands, 1936). Declining litter sizes in ageing common shrews (*Sorex araneus*) can be attributed to increased embryonic mortality (Brambell, 1935).

A number of species display a lowering of litter size with increasing population density. This has been recorded in lagomorphs (*Oryctolagus, Sylvilagus*) (Lloyd, 1963, 1964; Conaway *et al.*, 1960), and in the house mouse (*Mus*) where mean litter sizes range from 6.23 at low population densities to 5.11 at high densities (Southwick, 1958).

While these examples highlight the considerable intraspecific flexibility in litter size that exists in mammals, they do not always identify precise causes of the changes. Furthermore, more than one factor may operate at a time and ascertaining the role of each requires detailed analysis. For example, in the case of the bank vole in Czechoslovakia, litter size increased with parental weight but a relationship with parental age is less evident. However the most marked change in litter size was with season and it was therefore inferred that seasonal factors were more important than parental characteristics (Zejda, 1966).

3.4 Factors controlling breeding

It is common knowledge that most temperate mammals display a seasonality in breeding with the birth of young frequently commencing in late winter or spring and, according to the species, continuing for varying lengths of time into late summer or early autumn. The selective benefits of such a system are fairly obvious. The young appear when food supplies are increasing or at very adequate levels, the climate is favourable and protection is often afforded by good cover from the growing vegetation. However, many species are apparently capable of lengthening or shortening the breeding period from year to year. For example, in 1957 the breeding season of the rabbit in North Wales lasted from January to August, but in 1959 it extended only from February to June (Lloyd, 1964).

In the tropics the position is complicated because in some environments, such as rain forests, there are only slight seasonal variations in climate, while in others, such as the grasslands and woodlands, appreciable seasonal changes result from the seasonality of the rains. The latter, like the temperate regions, have favourable and adverse seasons.

In adapting to environmental conditions, some mammals have a greater opportunity for flexibility than others. Those with long gestation periods extending over many months (e.g. some deer, antelopes and marine mammals) may only have the opportunity to produce one or even fewer litters per year. Many of these animals experience a synchronization of breeding and a limited period of births. In contrast, many small mammals have all the reproductive characters (short gestation, post-partum oestrus, polyoestrus, rapid maturation) that favour repeated reproduction in a few months, and as a result offer greater opportunity for modification of length of breeding period. Even though these differences exist there must be one or more factors that determine when ovulation and spermatogenesis start and finish. It is the physical and biological factors implicated in the onset and termination of breeding that now merit further attention.

Within the geographical range of a species, the length or time of onset of breeding may alter with changing latitude. The North American deer *Odocoileus hemionus* breeds earlier in the winter further north, with mating occurring from late October to early November in Alberta, and from mid-December to mid-January in Arizona (Einarson, 1956). Within Britain, *Clethrionomys glareolus* has a longer and less intensive season in southern England and Wales than it does in northern Scotland (Brambell and Rowland, 1936; Delany and Bishop, 1960). Experiments on the effects of daylength or photoperiod have indicated a clear relationship in some species. In the common vole *Microtus arvalis*, studied in Poland, animals captured at the September equinox were subjected to gradually increasing photoperiod as would occur between 21 March and 24 June. By 23 October these animals were displaying sexual activity, while over the same period control animals under natural lighting were in anoestrus (Lecyk, 1962). The duration of the copulatory period in the male rock hyrax (*Procavia capensis*) in South Africa increases as latitude decreases. As no correlation has been found with rainfall and temperature it is assumed that photoperiod is important (Glover and Sale, 1968). There are many other examples, some inferred, of light controlling the time of breeding. This stimulus can be the result of shortening day length, with breeding commencing after the summer solstice (e.g. deer), or increasing day length with breeding commencing after the winter solstice (e.g. voles). These are referred to as *short-day* and *long-day* species respectively.

Temperature may play a significant role in stimulating breeding. Domestic sheep, kept captive throughout the summer months at autumn temperatures, come into oestrus seven weeks in advance of sheep kept at the same time at normal summer temperatures (Dutt and Bush, 1955). A positive correlation has been obtained between the date of first conception of European hares (*Lepus europaeus*) in New Zealand and the temperature in the two months preceding conception (Flux, 1967).

An adequate supply of appropriate foodstuffs is probably essential for successful breeding, and the natural provision of this food is often dependent on adequate rainfall and suitable temperatures. There is a threshold required for initiation, and intensity (i.e. the numbers of litters produced) can be related to the adequacy of the supply. Experiments with caged white-tailed deer (*Odocoileus virginianus*) on low and high nutrient diets have demonstrated how the latter came into oestrus earlier (Verme, 1965). In Australia, natural populations of rabbits have shorter breeding periods in poorly vegetated areas than in well vegetated areas (Myers, 1966). The provision of additional food in the spring to wild populations of wood mice advances the breeding season (Flowerdew, 1972). In the tropics, the breeding of the Nile rat (*Arvicanthis niloticus*) coincides with a change from grass and weeds to seeds and cereals as the main food—the production of the latter follows the seasonal rainfall (Taylor and Green, 1976). There can be delays in the manifestation of the effects of years of good and poor food supply. Thomson's gazelle (*Gazella thomsoni*), hartebeest (*Alcelaphus buselaphus*) and giraffe (*Giraffa camelopardalis*) living in East African bush-savanna, experience years of relatively high and

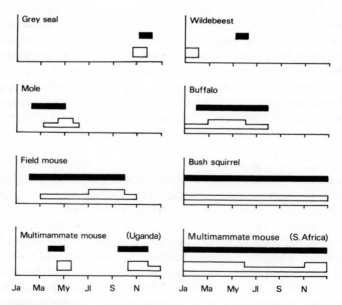

Figure 3.3 Periods of mating (solid blocks) and parturition (open blocks) in selected mammals. Peak periods of births are shown by raised blocks. Grey seal (*Halichoerus grypus*) (from Hewer, 1974), wildebeest (*Connochaetes taurinus*) (from Watson, 1969), mole (*Talpa europea*) (from Godfrey and Crowcroft, 1960), African buffalo (*Syncerus caffer*) (from Sinclair, 1977), long-tailed field mouse (*Apodemus sylvaticus*) (from Corbet and Southern, 1977), Boehm's bush squirrel (*Paraxerus boehmi*), (from Rahm, 1970), multimammate mouse (*Praomys natalensis*) (from Delany and Neal, 1969; Coetzee, 1965).

low rainfall which control the amount of vegetation produced. In years of plentiful food the animals attain good condition and in the year following have high conception rates (Field and Blankenship, 1973).

The density of a population may influence the length of the breeding season. This occurs in the field vole (*Microtus pennsylvanicus*), whose population cycles with a three- to four-year periodicity. When populations are increasing from a trough to a peak, the breeding season is protracted, but when populations are at a peak it is appreciably shorter (Krebs and Myers, 1974). Following the reduction of rabbit populations as a result of myxomatosis in Britain in the 1950's the breeding season was extended from about 16 weeks to 22 weeks or more (Lloyd, 1963). As with food supply, population density can influence the intensity of breeding. In elephants (*Loxodonta*) the interval between the production of calves increased from four to five years at low densities to eight to nine years at high densities (Laws, 1969).

These determinants of breeding apparently function in two ways. Proximate factors, such as photoperiod, may fix the time when the season starts. Subsequently, factors such as food supply may come into play which determine the length of the season and such reproductive parameters as the numbers breeding, frequency of breeding and termination of breeding; these are the ultimate factors. However, this probably oversimplifies the situation, as some factors (e.g. climate and food supply) must interact with each other.

3.5 Seasonality of births

There is a wealth of information on breeding seasons and particularly the times when young are produced. Generalizations concerning length and time of breeding seasons are difficult to make, but it is convenient to recognize different mammals as having short, long and continuous seasons of births, as long as it is appreciated that these categories may overlap.

Mammals having their young over a short period of a few weeks usually have a short mating period. Thus, deer that rut, seals that establish breeding colonies and wildebeest (Fig. 3.3) that have their young synchronously fall within this pattern. Some mammals, such as the European mole (Fig. 3.3), have shorter gestation periods than do these larger mammals, but as moles typically produce only one litter their period of production is short. The temperate bats fall within this pattern, being monoestrous and producing their young once a year when food is at a maximum. In these animals, earlier mating is followed by delayed fertilization or development. Some rodents, such as the African multi-mammate mouse, have a brief period of production of young in the short wet season of Uganda (Fig. 3.3). However, in the more attenuated season in South Africa, this species produces young throughout the year. The European badger is an exception—births are synchronized but matings spread over a long period (p. 23).

Mammals having births over several months bring this about by one of two strategies. They either have long gestation and mating periods or a short gestation and repeated reproductions. Among the former are many large mammals, such as buffalo in Serengeti, where after a 345 day gestation period young are born from January to July. The African elephant (*Loxodonta*) is also a seasonal breeder with births attaining a peak in advance of the wettest months. However, like many large African mammals the length of the season may be extended or shortened in response to local conditions. Several large mammals breed throughout the year but have distinct peaks of activity. Amongst these are Burchell's zebra (*Equus burchelli*) in Ngorongoro, Tanzania (Klingel, 1969), lechwe (*Kobus lechwe*) in Kafue, Zambia (Robinette and Child, 1964), and Coke's hartebeest (*Alcelaphus busephalus*) in Nairobi, Kenya (Gosling, 1969). It is to small mammals that we must look for the alternative strategy of producing young over a long period. Here, rodents such as white-footed (*Peromyscus*) and wood (*Apodemus*) mice undergo repeated reproductions through the temperate spring and summer, although under propitious conditions they may occasionally extend their breeding throughout the year.

Mammals producing young in fairly constant numbers throughout the year are frequently found in the aseasonal tropical rain forests, and include fruit bats, sciurid, murid, cricetid and echimyid rodents and primates. Here, in several species of rodents, including some squirrels from Africa and Malaysia, a relatively low proportion (25–50%) of adult females is pregnant at any one time, so that whilst young are continuously produced throughout the year there must always be a considerable number of females that are reproductively inactive (Harrison, 1955; Rahm, 1970). Continuous reproduction is not limited to tropical rain forests. In some desert regions where rainfall is low, some Australian marsupials, e.g. *Sminthopsis* spp., breed throughout most of the year (Tyndale-Biscoe, 1971), as does the jerboa (*Jaculus jaculus*) in the Sahara. The last lives on a constantly available food supply in the form of subterranean bulbs and corms and as a result avoids the seasonality of the rainfall (Ghobrial and Hodieb, 1973).

In the foregoing examples reference is made to species populations where all the individuals at a particular location conform to the pattern described. There are, however, a few examples where within one locality different individuals of a species adopt different reproductive strategies. This has been recorded in the leaf-nosed bat *Hipposideros caffer* which roosts in caves in northern Gabon. In some caves bats produce their young in April while the remainder do so in November. As the females are monoestrous and the males probably have seasonal quiescence this would appear to result in reproductive isolation of bats in caves in close proximity to each other (Brosset, 1968).

3.6 *r* and *K* strategies

In adopting a particular reproductive strategy the mammal is responding to the evolutionary pressures that have imposed broad limitations upon it, and to the current environmental conditions under which it now lives. Even within a single locality, both may be subject to considerable variation. The identification of life histories as *r* or *K* strategies provides a useful comparative base. The *K*-strategy is typified by a low rate of increase, delayed reproduction, small litter size, iteroparity (repeated breeding), large body mass, high survival, long life span and relative population stability. Many large mammals e.g., ungulates, exemplify this strategy. The *r*-strategists reproduce at an early age, have small body mass, large litter size, possibly semelparity (single breeding), short life span, variable population density and uneven use of resources. Many small rodents conform to this pattern. *K*-strategists are better adapted to predictable and equable climates and *r*-strategists to unpredictable and fluctuating climates.

While the recognition of *r* and *K* strategies has its uses and is of general application, some species do not conform—for example, many bats are *K*-strategists, and others display flexibility along the *r–K* continuum. Intensive removal of duiker (*Sylvicapra*) in Zambia over two years resulted in a higher proportion of calves in the population, i.e. increased reproduction, and a lower age of maturity (Wilson and Roth, 1967). This represents a shift in the *r*-direction. A study of Nile rats (*Arvicanthis niloticus*) (Neal, 1981) has demonstrated how the extent of *r*-ness can vary under differing environmental conditions. The western Uganda populations where environmental variability and stress are low have been compared with those of central Kenya where, in this semi-arid region, environmental stress is high. In the former, animals are large, litter size intermediate to small, reproductive rate and potential reproduction intermediate and the animals old at first reproduction. In Kenya (the more *r*-strategists) body size is small, litter size, reproductive rate and potential production are high and the rats young at first reproduction.

CHAPTER FOUR

SOCIAL ORGANIZATION

Mammals display a considerable complexity and variety of social organization. Types of association range from the relatively modest coming together of animals for mating and mother-infant interaction up to weaning, to the highly structured assemblages such as those found in the African hunting dog pack and the olive baboon troop. As social grouping and organization has become more complex, so has the need to recognize individuals and their behavioural intentions. This has been achieved through the evolution of numerous methods of communication, such as pheromones, tactile behaviour, posturing, vocalization, secretions, display, and agonistic actions. These are the elements responsible for maintaining group cohesion, and although the ways in which they operate provide a fascinating study, their detailed analysis is not appropriate here. There is, however, one general and widespread feature found in many mammal groups. This is the social hierarchy which involves a system of dominance and subordinance whereby some members of the group exercise authority over others as a result of their status. This can apply to males and females, but a single hierarchy in which both sexes are involved is infrequent. Social groups often display considerable co-operation between their members and frequently synchronize activities such as feeding, moving, and resting.

The main driving force in the establishment of social groups in mammals was undoubtedly the "invention" of milk, which forces a close relationship between mother and offspring at least until weaning. Association after this could persist if it were to the advantage of both parent and young. If the father comes into permanent association with mother and progeny, the result is a change from the one-parent family to the family group. Further increases in size could follow with the establishment of the extended family and the large grouping of several adult males and females and their

35

accompanying young. In fact, as shall be seen, all these types of association exist. The picture is even more complex, as there also occur other types of groups including harems (where typically one male has consort with several females) and exclusively male and female groups. The primates are an order which well illustrates a wide range of associations ultimately derived from and based upon a familial pattern.

The classification of social groupings according to familial units provides a useful structural basis for identifying levels of social organization. However, groups have probably evolved in response to ecological pressures and represent functional adaptations to them. This can result in social organization of a species displaying seasonal variability. This is often witnessed in the annual breeding cycle with its three phases (mate selection and copulation, parturition and lactation, and non-breeding). A classification which takes account of temporal changes in social structure is given in Table 4.1. Among species having seasonal changes in their social systems are the black bear, coati, red deer and potto—these only come together at mating and at other times the sexes are separate. Both sexes of seals occur together for much of the year and some species form harems for mating. Males and females of bats and ungulates (e.g. elk (*Alces*), and caribou (*Rangifer*)), occur together over much of the year with the sexes separating at parturition and lactation. Monogamous pairs in the seasonally invariant system include beavers, some duikers and the white-handed gibbon. Many species with larger groupings have seasonally invariant social structures including the house mouse, banded mongoose, wolf and African buffalo. In all these examples the group of females is tended by one or more males with no major social changes taking place. Other types of stable social organization include those of the lion where males associate with female prides rather than tend them (section 4.6), the Uganda kob where the system remains unchanged with some, but not the same, males constantly holding territory and mating

Table 4.1 Classification of social groupings

A. *Seasonally variant social systems*
    (i) Sexes come together at mating
    (ii) Mixed sex groups for much of year
        (a) Aggregate into harems for mating
        (b) Segregate at parturition and lactation

B. *Seasonally invariant social systems*
    (i) Monogamous pairs
    (ii) Larger assemblages

with entering females, and wildebeest males which also constantly hold territory but only mate successfully for a limited period of the year (section 4.5). The last two examples are sometimes referred to as "passive harems", as here the female group moves into and leaves the male territory at will. This contrasts with the active harem where a group of females is constantly accompanied and defended by the same male(s).

4.1 Social spacing

Social spacing is the disposition of members of a species with respect to each other. This covers the types of land tenure shown by mammals and the way in which individuals within a group occupying a common area are dispersed with respect to each other. Most mammals, whether solitary or members of a group, have an area of ground which they regularly patrol and with which they are well acquainted. Here they come to know features such as sources of food and water, bolt holes and escape routes, resting and lying-up sites, and look-out positions. This area is the *home range* (Jewell, 1966). Within it there may be a smaller *core area* which is used more intensively than the rest. Core areas are commonly found in species having a regular resting site, such as rodents and lagomorphs that live in burrows, and primates with nocturnal sleeping sites. The *territory* is a defended area. Its boundaries are well defined to the occupant(s) and are frequently marked (by urine, faeces, secretions) so that they can be recognized by strangers. Defence is by aggressive behaviour which can be expressed through a wide repertoire of visual, auditory and olfactory signals. Fighting can take place in some species. Territories which change with time, e.g. those of migrating wildebeest (*Connochaetes*), are called *spatiotemporal territories*. Also within this category are those territories that are defended at certain times of the day or year or both. These include the rutting territories of male deer and (by inference as a result of seasonal decline in agonistic behaviour) the territories of several species of small rodent.

The boundaries of the home range and the territory may coincide. There is then no overlap between the home ranges of different individuals or groups. This is found in gibbons (*Hylobates*), mating territories of many artiodactyls, and in prairie dogs (*Cynomys*). Alternatively, the territory can be confined to part of the home range and then ranges are capable of overlap. This is widespread in primates, e.g. vervet (*Cercopithecus pygerythrus*), gorilla (*Gorilla*), chimpanzee (*Pan*), and several carnivores where the relatively large areas some species such as wolves (*Canis lupus*)

Table 4.2   Home ranges of some European mammals

Species	Home range (ha)	Approx. weight	Locality
Pygmy shrew (*Sorex minutus*)	0.05–0.19	3 g	dunes, Holland
Common shrew (*Sorex araneus*)	0.02–0.08	8 g	dunes, Holland
Bank vole (*Clethrionomys glareolus*)	0.20♂, 0.14♀	20 g	woodland, England
Wood mouse (*Apodemus sylvaticus*)	0.23♂, 0.18♀	20 g	woodland, England
Weasel (*Mustela nivalis*)	9–25♂, 7♀	115 g♂ 59 g♀	farmland, Scotland
Stoat (*Mustela erminea*)	29–40♂ 4–17♀	300 g♂ 200 g♀	Finland
Red fox (*Vulpes vulpes*)	250–500♂, 250–500♀	7 kg♂ 5.5 kg♀	rural mid-Wales
Roe deer (*Capreolus capreolus*)	7.4–15♂, 7.1♀	25 kg♂ 20 kg♀	Dorset, England
Red deer (*Cervus elaphus*)	800♂, 400♀	95 kg♂ 70 kg♀	Rhum, Scotland

have to cover limits their effective territorial patrolling. Finally, in the absence of territoriality there can be considerable home range overlap. This is found in groups of female red deer (*Cervus elaphus*), cow elephants (*Loxodonta*), and barbary macaques (*Macaca sylvanus*). Our present knowledge of the behaviour of many small mammals is insufficient to identify territoriality. On the other hand their ranges are relatively easily measured, so that while home range can be delineated the existence of territoriality is much more imponderable.

Home range generally increases with size of animal (Table 4.2) and according to whether it is a herbivore or carnivore (Fig. 4.1) with the latter having the larger ranges. There are, however, still further factors involved in determining the size of home range. Increases in population density can result in decreases in home range. The range has been likened to an elastic disc which maintains a centre, with the area around it expanding and contracting in response to changing density, as has been found in house shrews (*Suncus*), deer mice (*Peromyscus*) and rabbits (*Oryctolagus*). One outcome of density change can be an alteration in territorial relationships. At low population densities, the home ranges of the howling monkey (*Alouatta*) overlap, while at high densities home range and territory coincide. Home range may also vary in relation to resources. In different parts of Russia the range of the stoat (*Mustela*) may vary between 10 and

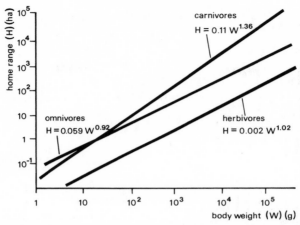

Figure 4.1 Relationships between home range and body weight in carnivores, omnivores and herbivores (modified from Harestad and Burnell, 1979).

200 ha according to the availability of food. Quality of the resource can also determine degree of overlap. In the forest- and thicket-dwelling vervet monkey, animals living in optimal habitat have little overlap in their groups, which occupy ranges of 9–28 ha, while groups living in marginal habitat have up to 25% overlap and occupy areas of 53–58 ha. Seasonal change in home ranges is often associated with sexual activity. In the male bank vole, increased range size takes place during the breeding season, at which time the range of a single male comes to overlap those of several females (Moor and Skeffens, 1972). Finally, there can be permanent shifts in home range as has been recorded for several small rodents. This gradual permanent change should not be confused with the ranges of young, transient animals which are frequently seeking a permanent range.

The pattern of distribution within a group is often significant. Where there is a dominant individual, as in groups of red deer hinds and calves, this animal may provide the leadership and situate itself at the head of the group. In this species the second in the matriarchy brings up the rear. In some large bisexual groups of foragers, e.g. olive baboon (*Papio anubis*), there is a well-defined disposition of adults and young (section 4.7) which serves for protection and food location. Group dispersion in cetaceans is principally related to navigation and feeding (Fig. 4.2).

Among the larger terrestrial mammals are two common and interrelated functions of group configuration. These are concerned with anti-predator and predatory behaviour. Defence from attack by the spotted hyaena

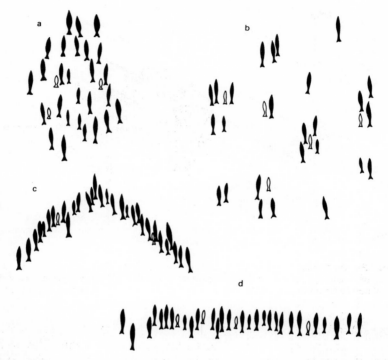

Figure 4.2   Herd configuration of northern right whale dolphins (*Lissodelphis borealis*). (*a*), tightly packed groups; (*b*), groups divided into sub-groups; (*c*), V-formation; (*d*), chorus line (modified from Leatherwood and Walker, 1979).

(*Crocuta crocuta*) varies from species to species (Kruuk, 1972). Zebra live in units comprising one stallion and several mares and foals. When attacked by a clan of hyaena, the females and young form together in a tight group and trot away at a steady pace. Meanwhile the stallion keeps a short distance behind, chasing, biting and kicking the hyaenas (Fig. 4.3). Stallion and mares are never separated by a large distance and if the stallion falls behind, the mares slow their pace. The hyaena strategy is to sustain the attack from behind on stallion, mares and young. Eventually success can be attained by their plucking out of the group a young animal which the clan then sets upon. Their success is minimized by the aggression of the stallion, close packing of the female group and the steady trot away from the attackers. When hyaenas approach eland (*Taurotragus oryx*) the defence consists not of running away; instead, the adult cows stand their

Figure 4.3 Anti-predator behaviour in zebra (*Equus burchelli*). The stallion is at the rear fending off hyaenas (*Crocuta crocuta*) while mares and foals move away together in a tight group (modified from Kruuk, 1972).

ground, forming themselves into two groups (Fig. 4.4), one which surrounds the calves, the other facing the hyaenas in a line. The latter actively attacks the hyaenas. Under these circumstances the hyaenas have difficulty launching a successful attack.

4.2 Social organization in marsupials

Social organization is generally poorly developed within the marsupials. The American opossums (Didelphidae), the marsupial "cats", "mice" and "rats" (Dasyuridae) and the wombats (Phascolomyidae) are solitary for most of their lives. Within the Phalangeridae there is limited grouping around the family, as in the glider (*Petaurus*) which has a male-dominated extended family. It is in the kangaroos and their allies (Macropodidae) that the greatest development of sociality is found, although even here it is not highly organized.

One such example is the whiptail wallaby (*Macropus parryi*). (Kaufman, 1974). This small wallaby occupies *Eucalyptus* woodland with a generous grass cover on which it feeds, supplementing the diet with herbaceous plants. The whiptail wallaby organizes itself into stable "mobs" of 30–50

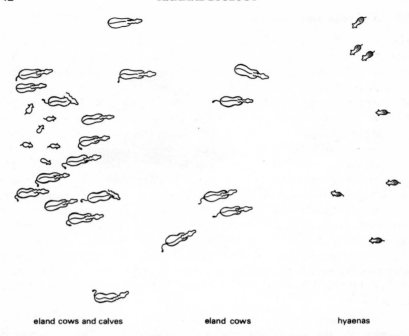

eland cows and calves   eland cows   hyaenas

Figure 4.4   Anti-predator behaviour in eland (*Taurotragus oryx*). One group of cows forms a front line facing the hyaenas while other cows group around the calves (modified from Kruuk, 1972).

individuals. Each mob occupies a range of 70–110 ha in which there is limited overlap with adjacent ranges. When mobs meet on range boundaries, there is little or no expression of aggressive behaviour—the members of the two mobs intermingle, rest and feed together. On separating, the mobs retain their original identity, with little interchange of members taking place. At night, members of a mob aggregate under trees, and by day subgroups forage over open ground. During this time these subgroups may coalesce, exchange members and break up again before finally reassembling for the night. The wallaby mob is a weakly organized society which for short periods appears unstructured. Adult and sub-adult males have a linear dominance hierarchy although no such hierarchy has been found in females. Aggressive behaviour and the assertion of dominance is largely ritualized, and even the most aggressive encounters seldom result in physical harm to the contestants. Dominant males have ready access to oestrous females.

4.3 Social organization in bats

The availability of food and roosting sites are the most important determinants of bat social structure. As small mammals, bats are confronted with problems such as vulnerability to predators, loss of heat (particularly in cold weather) and a need for relatively high food intake. Many species avoid the adversities of harsh climate by hibernation within a protective roost; the roost must also perform this function during the diurnal period of inactivity. Inevitably, the capacity of the roost, together with the availability of food within foraging distance, must determine the numbers that can be supported. Among the bats there is considerable variation in the preferred type of roost. Thus, while caves are large and provide excellent cover they are widely dispersed; in contrast, hollow trees, overhangs, the undersides of leaves and tree tops are widespread and more abundant. Those species utilizing small roosts can be expected to live in smaller groups, but by virtue of their abundance the bats using them are able to exploit extensive areas of habitat. Nevertheless, they are more vulnerable to predators and climate, and it is of note that only the larger species of tropical bats (*Pteropus, Eidolon*) form large colonies in exposed situations.

A wide range of bat social structures (Bradbury, 1977) has developed in response to the differential application of these various ecological pressures.

Solitary except for copulation and mother-offspring associations

Few bats are solitary, and of these, most are tropical. Representatives exist among the Megachiroptera (e.g. *Epomops franqueti, Megaloglossus woermanni*) and Microchiroptera (e.g. *Diclidurus alba, Pipistrellus nanus*). Some of these tropical species breed throughout the year.

Sexes separate between copulation and parturition

In several species the sexes separate after mating and come together again after varying lengths of time subsequent to the birth of the young. Among those bats spending most of the year in sexually segregated groups are species from the West Indies (*Pteronotus parnelli*) and Sri Lanka (*Hipposideros atratus*). In contrast, in many species of fruit bats (Pteropidae) the sexes associate for longer periods. In the Australian *Pteropus poliocephalus*, males join females in spring soon after they give birth. The males set up territories around the females, with whom they remain until immediately after copulation in late summer. Here the sexes are apart for approximately eight months.

Sexes segregate at parturition and are together at other times

This category embraces what is sometimes called the "temperate cycle", in which mixed sex groups occur during the non-breeding period. The pattern is typical of temperate bats and as a result includes many species from several families. A typical example is the noctule (*Nyctalus noctula*) which is widespread through the northern Palaearctic. The hibernating groups include both sexes. In spring the females collectively give birth to their young in so-called "nurseries" from which the adults go out in search of food. Nursery groups are not static, with interchange commonly taking place. At this time the males are dispersed and solitary. In late summer the females leave the nurseries, and move to transient harems for mating. After a short time these are vacated, and both sexes assemble at the overwintering hibernation sites.

Variations have been found in other species with respect to the winter, parturition and mating groups. In winter groups the disposition of young and adults, sexual distribution and sex ratios are among the variable characters. In the greater horseshoe bat (*Rhinolophus ferrum-equinum*) adults are separated and juveniles clustered, a possible result of different physiological requirements of large and small animals. There is a sexual separation in colonies of the serotine *Eptesicus fuscus* (the males occupy the cooler areas) while in the lesser horseshoe bat (*Rhinolophus hipposideros*) the proportion of males increases through the winter. At parturition both sexes of the tropical *Taphozous melanopagon* are present, the males forming a ring around the centrally grouped females. In several species (e.g. *Rhinolophus lepidus, Myotis natteri*) there is a local segregation of sexes at the same site. In contrast, some nurseries (e.g. in *Taphozous nudiventris*) accommodate adult males and females at the time of parturition. Social organization of parturition groups can vary within the species. Males of the long-eared bat (*Plecotus auritus*) are present in small numbers at parturition in British colonies but are absent from French and Russian nurseries. Sexual segregation in nurseries can be terminated by males entering the group (e.g. *Myotis myotis*), females leaving the young at weaning (e.g. *Pipistrellus pipistrellus*) and the young leaving the adult females (e.g. *Chalinolobus dwyeri*). Little is known of mating group organization. In *Myotis sodalis*, mating lasts for one to two weeks in autumn. The sexes roost separately in the hibernation site and at night females enter the male chamber for copulation. It is possible that transient mating sites can occur along a migration route (e.g. *Tadarida*).

Year-round harems

These have been recorded for several species of tropical bats. One example is the large, omnivorous, cave-dwelling, neotropical spear-nosed bat *Phyllostomus hastalus*. The harem consists of 10–100 females per male with the remaining males forming loosely structured groups of 20–50 animals. The harem males defend their females from intruding males and maintain a stable social structure with little turnover in harem composition. The young are born over a limited period and remain in the colony for three to five months. The sheath-tailed bat *Saccopteryx bilineata* from Trinidad maintains a year-round harem in which the groups are particularly unstable. The harem comprises one to eight females, and there is continuous and intense competition among the males for possession. Furthermore, some females move from harem to harem even though there is appreciable aggression between them. This pattern of social structure, like the two that follow, is seasonally invariant.

Year-round multi-male/multi-female groups

The large fruit bats *Pteropus giganteus* of India and Sri Lanka and *Eidolon helvum* of Africa live in large mixed-sex colonies throughout the year. Parturition takes place over a limited period, the young being born into these large and sometimes enormous colonies. There is little evidence of social organization or the development of hierarchy. The African fruit bat may make local seasonal migrations.

Monogamous families

At present it is only possible to infer the existence of monogamous units from the occurrence of species which are commonly seen in pairs with or without young. Examples include the yellow-winged bat (*Lavia frons*) and woolly bats (*Kerivoula*) from Africa, the leaf-nosed bat *Hipposideros brachyotis* from Sri Lanka, and the carnivorous *Vampyrum spectrum* from tropical America.

4.4 Social organization in rodents

In spite of their abundance and multiplicity of species, social structures within wild rodents have been little studied. This is probably because their small size and secretive habits make direct observation difficult. Apparently, many species are solitary for much of their lives, such as some sciurids, pocket gophers (Geomyidae), kangaroo rats (Heteromyidae), mountain beavers (*Aplodontia*), many murids, and jerboas (Dipodidae).

A slightly higher level of organization is found in the beaver (*Castor*), where the family forms the social group. There is considerable mutual tolerance and no apparent social integration.

The house mouse (*Mus musculus*) frequently occurs in family groups, typically of two or three males and four or five females and their young. There is a dominant male who is larger and older than the subordinates. He patrols the territory which is defended against intruders by all members of the group. The social order is stable within the group, with little overt expression of aggression. Most fighting takes place between males at territorial boundaries. In experimental high-density populations (Crowcroft and Rowe, 1963) larger groupings can occur, but they are still dominated by a single male who usually nests alone or with an adult female. The remaining males, according to their status, occur as individuals dispersed through the colony, in cowed nesting groups or taking cover among females and juveniles. Under natural conditions the younger males would probably be forced to disperse. In the Canadian prairies (Anderson, 1961), small stable groups were found to live in the favourable conditions provided by grain stores, while fluctuating large assemblages, found in less hospitable open country, were the excess produced by the former. Mice on the island of Skokolm off the Welsh coast behave in the same way.

This extended family social structure probably occurs in other species of small rodent, including the Palaearctic woodmouse (*Apodemus*) (Brown, 1969) and the North American black-tail prairie dog (*Cynomys ludovicianus*) (King, 1955; Smythe, 1970; Smith *et al.*, 1973). In the latter, groups (or "coteries") have their burrow systems juxtaposed to form a "town" or "colony", frequently of many thousand individuals. The members of a coterie average 1.65 adult males, 2.45 adult females and 5.93 immatures. With increase in size, fission takes place to form smaller groups. The coterie occupies a defended territory whose limits remain virtually unaltered from generation to generation. This is quite extraordinary as the rate of population turnover is rapid and emigration frequent. New coteries are established by the departure of adult males to new locations where they commence burrowing. They are subsequently joined by a few adult females, and it is the subadults and juveniles left behind in the old burrows that organize and retain the new coterie within its original boundaries. As the prairie dog inhabits open, exposed places it is particularly vulnerable to predators. Its protection is afforded by a characteristic vocalization (described as a "high-pitched nasal yipping") which, once initiated, spreads in a wave across the whole town.

A unique level of mammal social organization has been found in the naked mole rat (*Heterocephalus glaber*) (Jarvis, 1981) which inhabits the drier, warmer parts of eastern Africa. This almost hairless rodent is exclusively subterranean, lives in colonies of up to 40 (and possibly more) individuals and inhabits an extensive network of foraging tunnels. Within the colony are recognizable "castes" comprising frequent workers (Fig. 4.5), infrequent workers, non-workers and a single breeding female. The frequent workers are the smallest members of the colony and are highly active. They dig, transport soil, forage and carry food to the communal nest. The infrequent workers are larger than the frequent workers and carry out the colony tasks at a much slower rate than the frequent workers. The non-workers are the largest members of the colony. They spend much time huddling in the nest and assist with the care of the young. The social role of these animals has not been clearly identified

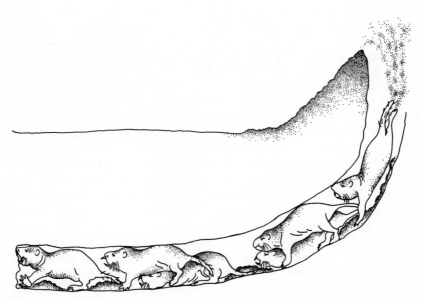

Figure 4.5 Co-operative digging by "frequent workers" within colony of naked mole-rat (*Heterocephalus glaber*) (modified from Jarvis and Sale, 1971).

although it is probable that the breeding males come from this group. In all these castes the females are non-breeding, with quiescent ovaries. All males are potentially able to breed. The breeding female is of similar size to the non-workers and produces 1–4 litters a year of up to 12 young. On the removal of the breeding female, another female often becomes reproductively active. At times of danger, young are removed from the communal nest, where they are tended by all castes, by the frequent and infrequent workers. The young grow slowly and initially enter the frequent worker caste. Some individuals grow faster than others and eventually become non-workers and possibly breeding females. This social structure is suggestive of an extended family with considerable co-operative breeding.

4.5 Social organization in African artiodactyls

The antelopes of the African savanna probably evolved from forest stocks that moved into the grasslands. Forest antelopes are generally small and solitary, but once in the savanna their social and ecological evolution took on new dimensions. They radiated to produce a large number of species (African antelopes comprise 37% of the world ungulate fauna), many became larger than their antecedents, and many organized themselves into social groups which afforded greater protection in this more open habitat. The abundance of large species in grassland can be explained ecologically. In the seasonally wet and dry savanna, the vegetation, particularly the grasses, fluctuates in quality and quantity. Small antelopes have relatively greater metabolic requirements than large ones and as a result consume foods of high energetic value. Their small size permits the delicate selection of preferred foods. As these items are relatively uncommon, the small antelopes are usually dispersed and at relatively low densities. Large antelopes consume considerable bulk of lower quality food which they utilize relatively efficiently. These factors are taken into account in adopting an ecological and behavioural classification of five levels of social organization (Table 4.3) (Jarman, 1974). Here, the family and its extensions once again provides a social focus.

The lesser kudu (*Tragelaphus imberbis*) is an example from Class B. This antelope with its characteristic spirally twisted horns inhabits bush and thicket in eastern Africa. Adult males are solitary and territorial. Females live in groups of 2–10 comprising several adults and accompanying immatures of both sexes. The small group size and the permanent holding of territory by males distinguishes this species from those in Group C such

Table 4.3 Social organization and feeding behaviour of African artiodactyls (after Jarman, 1974)

Social organization	Group size	Weight kg	Feeding selectivity	Examples
CLASS A				
Single pairs or pairs with offspring	1–3	1–20	Food spatially scattered; selective from wide range	Dik-dik, duikers, klipspringer, grysbok
CLASS B				
Several female–offspring associated. Permanent home range within range of single males	Usually 3–6	15–100	Feed on range of grass or browse selecting particular parts	Reedbuck, rhebuck, oribi, lesser kudu
CLASS C				
Larger herds. In breeding season a few males hold territories and exclude other males (solitary males, bachelor herds)	6–several 100 (+solitary ♂♂)	20–200	Feed on variety of grass and browse with wide dispersion. Selective on parts of plants	Kob, waterbuck, puku, springbok, gazelles, impala
CLASS D				
As class C for sedentary periods; during seasonal migration herds coalesce into superherds	6–several 1000 (+solitary ♂♂)	100–250	Grass but selective of parts. Seasonal patchiness results in migration	Wildebeest, hartebeest topi, blesbok
CLASS E				
Large herds; ♀♀ and young and many adult ♂♂ with dominance hierarchy. Bachelor herds occur	Several 100 up to 2000	200–700	Unselective. Grasses and browse often of low nutritive value	Buffalo, oryx, gemsbok

as the Uganda kob (*Kobus kob*) which lives on grassy plains in Uganda and Sudan. In the kob, sexes are segregated when not breeding, with males in non-territorial bachelor herds of up to 500 individuals and females typically in groups of 30 to 50. Herd size is partly determined by food availability. The males form territorial breeding grounds or "leks" in which there is a group of about 15 small central territories and a further 25 larger peripheral territories. As breeding takes place throughout the year the territories are permanent, although ownership changes frequently. The smaller central territories where most mating takes place have the shortest occupancy.

The social organization of the blue wildebeest (Class D) includes nursery groups of about ten adult females and young, single male territories and bachelor herds without territories. The single males primarily defend their territories throughout the year for courtship, even though rutting takes place over a limited period. With this organization the male defends the nursery group for however short or long a period it chooses to remain in the territory. During migration the males either defend an area for a few hours to a few days, or they have a moving zone about themselves unrelated to any particular area. The great migrating superherds are formed from the coalescence of the three social components, with each retaining its identity through the migration.

The large gregarious African buffalo (*Syncerus caffer*) (Class E) lives in herds of 50–1000 individuals containing many adult males and a constant female membership. There is no territoriality, although herds occupy recognizable home ranges. The adult males form linear hierarchies, with high status conferring mating preference. Sub-adult males often form sub-groups within the herd while old males frequently leave the herd to live as solitary individuals or form bachelor groups. This example contrasts with those in Class B to D in that the animals are not territorial, there is little aggression between the males and there are permanent mixed-sex groups.

Defence against predators in Class A and B takes the form of freezing in cover or fleeing, while members of Class D and E are prepared to stand ground and defend against attackers. Class C animals may freeze or run; in the latter case the large herd disperses and subsequently reassembles.

4.6 Social organization in Carnivora and Pinnipedia

The carnivores display a range of social organization from the mainly solitary individual to the complex society. These do not relate to phylogenetic division—indeed, in some instances the simplest and most advanced are found in the same genus, e.g. *Felis, Canis*. Solitary carnivores typically feed on prey items smaller than themselves, but association into organized groups makes possible the consumption of appreciably larger animals. As a result of their feeding habits their home ranges are relatively large; in many cases, e.g. polar bears (*Thalarctos*), their territories are too large to be defended simultaneously. Among the solitary carnivores are some cats, dogs, stoats, otters, bears and mongooses, but inevitably association occurs at mating and when the young are being reared.

An unusual example of the solitary territory is found in the American black bear (*Ursus americanus*). The adult female assumes a territory which she vigorously defends during the mating season and which covers 10–25 km². This territory she subsequently assigns to her female offspring, permitting them to establish themselves in sectors of it. They ultimately take over her sector when she dies or leaves. Females that only manage to obtain a small territory do not breed. The males are not involved in this system and leave the maternal association as sub-adults. Adult males display agonistic behaviour towards each other during the mating season but thereafter assemble in peaceful feeding aggregations until returning to the female territories where they lie up in dens for the winter.

A more advanced association is found in the coati (*Nasua narica*) of central America (Kaufman, 1962). One to four adult females and their young live gregariously in bands of 4–13 animals. This loosely integrated group can also include adult females without young. Each band has a home range with a core area (15–20 ha) occupied for approximately 80% of the time. Ranges overlap but not core areas. When two bands approach they show mutual avoidance rather than aggression. The solitary male has a territory overlapping that of the females, but association with them only occurs for about a month of the mating season. At all times he is subordinate to the females. There is poor social organization within the groups, with no evidence of leadership or group hunting, and all animals are markedly individualistic. These coati groups probably arise as extended families.

A more elaborate social unit, probably also deriving from the extended family, is found in the banded mongoose (*Mungos mungo*), a common species of the African savanna (Rood, 1975). Packs contain 2–29 individuals, including adults of both sexes and young. Their ranges overlap, and when two packs meet they exchange vocalizations, chase each other, mark with secretions, fight and finally disperse. There is no defined hierarchy within the pack, which inhabits a den at night. When out foraging in daytime, they disperse themselves within sight and sound of each other. When food (frequently insects) is located, vocalization by the finder attracts others to it. The pack sometimes drives a small predator away from its prey. In addition to these feeding advantages, group membership provides better protection from predators. By clustering and displaying aggressive behaviour, or by disruption resulting from rapid dispersal in several directions, they can often escape predators. The mongoose pack is a well-organized society with social interactions between its members such as grooming, mutual care of young, male guarding of young when the pack is away foraging, vocal communications (chirrups, twitters, grunts) and occupation of a common den.

Among seals, the establishment of a breeding harem is common. In the grey seal (*Halichoerus grypus*) of the northern Atlantic, cows come ashore to pup in late autumn and early winter (Hewer, 1974). Within about two weeks of giving birth they come into oestrus and are then ready to mate. In the meantime the bulls have set up territories along and above the shoreline of the isolated and often remote coasts where breeding occurs. Larger bulls have priority and all have to vigorously defend their territories throughout the three to four weeks of the mating period. Each rookery contains from 7 to 13 cows. Mating takes place several times between the bull and each cow and as a result exhaustion and replacement of the bull can occur. Cow replacement is also possible. As cows come ashore in the first instance to pup, speculation exists as to where virgin cows are first mated. There are surplus young males which at this time of year are scattered about rock exposures and small islands, and these may then encounter young females with which they mate, but this is unconfirmed. Outside the breeding season it is assumed that the two sexes intermingle freely in their marine environment.

It is in the lion, wolf and hunting dog that the most complex carnivore social organizations are found. In the lion (*Panthera leo*) the pride, usually of 10–20 adults and sub-adults, is essentially female-dominated (Bertram, 1975). All the adult females have a close female relative (mother, daughter, sister) within the pride, which defends a territory against other female prides. The males have a loose association with the female pride, but form their own prides which also occupy defended territories. Males associate with females for feeding. Adult lionesses maintain an optimal pride size by driving out excess young sub-adult females, which become nomadic vagrants with little expectation of joining another pride. In defending their territories, males are also defending their female prides. Males will also drive out young members of their own sex. There is much fighting between males for membership of a pride and the females that go with it—only fully mature males succeed, and when established leave the females to stalk and capture prey. Cubs born soon after a new male takes over a pride are usually killed by it. The female then comes into oestrus and mates with the new pride male whose genes are then added to the population. This remarkably complex social structure produces many advantages. It regulates pride size to food availability, it ensures more efficient hunting and it confers certain social benefits such as communal suckling, male tolerance to cubs at kills and lack of competition between males for oestrous females.

The wolf (*Canis lupus*) is distributed across much of North America and Eurasia, although its range has been steadily reduced over several centuries as a result of man's activities. This species hunts in packs of typically 5–15 animals (Mech, 1970). Packs are constantly on the move in search of food, some following migratory herds (e.g. caribou) in their winter and summer movements. Home ranges per individual in North America vary between 25 km^2 in Algonquin Park, Ontario, to 168 km^2 in Michigan. During the denning period when young are being reared, ranges are temporarily reduced. Wolf packs stalk, chase and kill their prey, which is frequently relatively large, e.g. moose, caribou, or elk (*Cervus canadensis*), as a co-ordinated unit. They often drive prey and simultaneously attack it from several angles. Within the packs there is separate linear dominance of males and females—the dominant male is the leader in chases for food, responds first to intruders and has mating priority. Rank is sustained through a panoply of signals including facial expressions, barks, howls, tail movements and postures of the body and occasionally fighting that can be violent. New packs can be formed when a male and pregnant female break away from an existing pack. Territory marking is achieved by urine and secretions from the anal glands, and packs avoid areas for some time after they have been visited by another pack. If two packs happen to meet, fighting breaks out between them and fatalities can result, but usually and not surprisingly the larger pack is successful and the smaller pack moves away.

Like the wolf, the African hunting dog (*Lycaon pictus*) hunts in packs (Lawick and Lawick–Goodall, 1971; Lawick, 1974). Here too herbivores that are larger than the dog are frequently taken. The leader of the pack selects the prey, which is then pursued and attacked by the pack, of up to 21 animals. On the open African plains where these animals live, Grant's and Thomson's gazelles (*Gazella* spp.), impala, calves and adults of the wildebeest and zebra are among the common prey items. The hunting dog has a higher success rate in capturing its prey than the wolf as it is seldom intimidated or deterred by the prey. Once one animal seizes the prey, the remainder join in the attack. Within the pack are separate male and female hierarchies. These are not easily recognized, as much of the behaviour has become highly ritualized and packs are peaceful, placid associations. The high degree of sociality is seen in the feeding behaviour of the pack towards females with young, old animals and pups too young to indulge in hunting. The pack overeats on the kill and on its return to the den its members regurgitate food for the animals left behind (Kühme, 1965). A further subtle adaptation is found in their reproduction. While the young are being fed by the mother, the pack loses some of its mobility and it has to regularly return to the same den. The synchronization of breeding minimizes this, and this has been achieved in a most intriguing way. Only one or two adult females in the pack breed each year. However, their litters are extremely large (up to 12), thus affording a high replacement rate; furthermore, the birth of all the individuals in a single litter is obviously synchronized. This mechanism contrasts with the typical canid situation, where all adult females would be producing smaller litters with less synchronization, the gross food requirements of the hunting pack would be greater, while its hunting capacity would be reduced because proportionately more females would remain in the den. The hunting dog ensures that large numbers of litters are not reared by the active killing of pups from low-ranking females. As these animals cover such vast areas there are few contacts with other packs, and the staking of territorial boundaries is not of great importance—the only exception is in the vicinity of the den.

4.7 Social organization in primates

As in other mammal orders, there are several levels of social organization. At its simplest are the solitary species, such as the slow-moving, arboreal potto (*Perodicticus potto*) of the African rain forest. This animal (which feeds on fruits, gums, resins and insects) is territorial—the male and female both have exclusive territories of approximately 12 ha and 7.5 ha respectively, although male and female territories do overlap. Territorial boundaries are marked by urine; vocalization is rare. Even at courtship and mating there is minimal association, the young leave the mother soon after weaning, and the solitary existence is re-established.

The white-handed gibbon (*Hylobates lar*) is a small, arboreal pongid inhabiting the rain forests of south-east Asia. This primate lives in small stable family groups comprising a monogamous pair of adults and up to four offspring (Carpenter, 1940; Ellefson, 1968). Occasionally an ageing male may be present within the group. Solitary individuals also occur infrequently. The family groups live in territories of 100 to 120 ha, which are defended equally by the adult male and female. Within the group, dominance is weak or absent, with the members of the family remaining close together at all times. Both parents involve themselves in rearing the infants and while the female typically carries the young and has the closest association, the male also plays with, grooms and on occasion carries the young. These gibbons feed on fruits and foliage, spending much of their lives in the upper canopy. Occasionally they descend to the shrub layer and to streams to drink water. They are distinctly vocal, having a range of calls and grunts emitted in defence of territory and intragroup behaviour.

The savanna-dwelling olive baboon (*Papio cynocephalus*) lives in multi-male troops of 12–185 individuals (Hall and De Vore, 1965). In this species, males are appreciably larger than females. The troop is organized about a dominant male who leads the hierarchy by his aggression and ability to enlist the support of others. Even so, there is appreciable male tolerance with several adult males living amicably within the troop. The females display little dominance. These baboons are semi-terrestrial, feeding on a variety of seeds, fruits, bulbs and insects. As they forage over the ground they are dispersed in a regular fashion so that the dominant and large males are centrally located, the females with young close to them and the remaining members of the troop located more peripherally (Fig. 4.6). Subordinate males act as sentinels at the front and rear. On encountering a hazard or danger, a dominant male moves forward to investigate. Dominance confers mating priority so that most young are fathered by relatively few adults. Of considerable interest is the cohesion and integrity of these large multi-male groups. This is made possible by the numerous calls and visual and tactile signals that have evolved. The social system provides protection from predators, feeding efficiency and selective breeding.

In addition to these three levels of social organization, there are also examples of the uni-male troop, where one male is accompanied by several young males (who show submissive behaviour to the dominant adult male) and by females with their young (e.g. the howling monkey, *Alouatta palliata*). Here the adult male is intolerant of the young, which are forced to leave the troop and enter a bachelor group. Finally, there is the age-graded male troop in which the oldest male shows greater tolerance to younger males so that more than one adult male may be present. Here there is a linear hierarchy usually based on age. This is found in the vervet

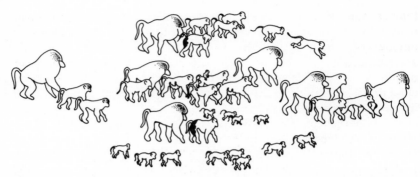

Figure 4.6 The positions of individual members within the troop of olive baboon (*Papio cynocephalus*) during movement. Females with young are in the centre and surrounded by adult males. Juveniles are round the periphery. Other adult males and females precede and follow the centre of the group. Two oestrous females, with dark hindquarters, are closely attended by adult males (modified from Hall and de Vore, 1965).

(*Ceropithecus pygerythrus*) and gorilla (*Gorilla*). Age-graded troops tend to be larger than uni-male troops and frequently split into sub-troops, which minimizes aggression between males. Primate groups can be further classified according to their feeding habits and favoured habitats. Feeding habits are· broadly divided into fruit-eating (frugivorous), foliage-feeding (folivorous) and insect-eating (insectivorous) while habits include the arboreal, terrestrial and semi-terrestrial (examples are given in Table 4.4). While many species have a fixed social structure, others, frequently inhabitating a range of habitats, are more flexible. Thus the grey langur (*Presbytis entellus*) forms uni-male troops in roadside forest remnants in India, whereas in more extensive areas of forest the age-graded troop is common.

Attempts have been made to draw broad ecological conclusions for all primates by relating such variables as group size, population density, feeding habits, times of activity, home range and size of animals. This is a particularly difficult analysis with such variable and flexible material. Nevertheless, it is a worthwhile exercise from which interesting conclusions are emerging (Clutton-Brock and Harvey, 1977). Nocturnal species are generally small, feed frequently and live alone or in pairs, typically inhabit the middle and lower strata of forest and depend on crypsis rather than flight or defence to avoid predation. They also occur at lower densities than the few species of similar size that are diurnal. In diurnal genera, body weight and feeding group size tend to be larger in terrestrial than arboreal species (Fig. 4.7). As might be expected, home range increases with group

Table 4.4 Examples of social organization and feeding ecology in primates (after Eisenberg *et al.*, 1972)

SOLITARY SPECIES	
A. Insectivore-frugivore	lesser mouse-lemur (*Microcebus murinus*)
	potto (*Perodicticus potto*)
B. Folivore	sportive lemur (*Lepilemur mustelinus*)
PARENTAL FAMILY	
A. Frugivore-insectivore	common marmoset (*Callithrix jacchus*)
	dusky titi (*Callicebus moloch*)
B. Folivore-frugivore	indri (*Indri indri*)
	white-handed gibbon (*Hylobates lar*)
UNI-MALE TROOP	
A. Arboreal folivore	mantled howler (*Alouatta villosa*)
	Abyssinian colobus (*Colobus guereza*)
	grey langur (*Presbytis entellus*)
B. Arboreal frugivore	white-throated capuchin (*Cebus capucinus*)
	blue monkey (*Cercopithecus mitis*)
C. Semiterrestrial frugivore	Patas monkey (*Erythrocebus patas*)
	gelada (*Theropithecus gelada*)
AGE-GRADED—MALE TROOP	
A. Arboreal folivore	Hanuman langur (*Presbytis entellus*)
	mantled howler (*Alouatta villosa*)
B. Arboreal frugivore	black-handed spider monkey (*Ateles geoffroyi*)
C. Semiterrestrial frugivore-omnivore	vervet monkey (*Cercopithecus pygerythrus*)
D. Terrestrial folivore-frugivore	gorilla (*Gorilla gorilla*)
MULTI-MALE TROOP	
A. Arboreal frugivore	(*Lemur fulvus*)
B. Semiterrestrial frugivore-omnivore	Japanese macaque (*Macaca fuscata*)
	olive baboon (*Papio cynocephalus*)
	chimpanzee (*Pan troglodytes*)

weight (this takes account of numbers and size). Frugivore groups require larger ranges than folivores (Fig. 4.8)—this could be because fruits are more thinly distributed than foliage. The extent of daily movements or day range is generally greater in terrestrial than arboreal species. Among the frugivores, increasing feeding group weight is associated with increasing day range. No such correlation exists for the folivores. Finally, there is a negative correlation between individual body weight and population

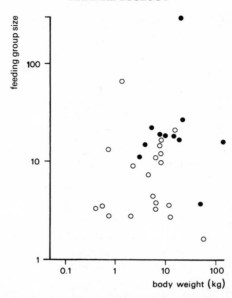

Figure 4.7   Feeding group size of terrestrial (●) and arboreal (○) diel-active primate species in relation to body weight (modified from Clutton-Brock and Harvey, 1977).

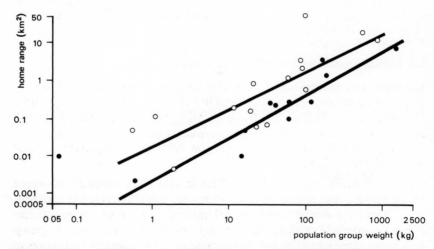

Figure 4.8   Home range size in relation to population group weight for different species of primates. The upper regression line is for frugivores (○) and the lower one folivores (●) (modified from Clutton-Brock and Harvey, 1977).

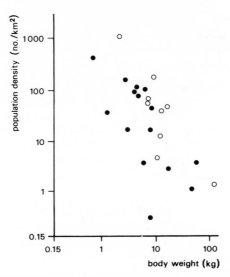

Figure 4.9 Population density in relation to body weight in different species of primates—folivores ○, frugivores ● (modified from Clutton-Brock and Harvey, 1977).

density with the diurnal folivores occurring at higher densities than the frugivores (Fig. 4.9). However, the similarity of the slopes suggests that with increasing body size, relatively the same amount of food is available to those two feeding groups.

4.8 Ecological costs and benefits of being social

The ecological evaluation of social organization considers aspects such as its effects on resource utilization and intraspecific population regulation. The latter encompasses changes in perinatal mortality, breeding success and survivorship, all of which can be influenced by the social structure of the species concerned. Social organization can result in improved survival in the following ways:

(i) *Protection from predators.* The existence of sentinels in some species, or simply a multiplicity of alert individuals, provides greater awareness of predator incursion. Further protection can be afforded by vocalization (e.g. monkeys, prairie dogs), group aggression toward the predator (e.g. olive baboons, eland cows), and disruption through explosive dispersion (e.g. banded mongoose).

(ii) *Improved feeding efficiency.* Searching capability increases because food found becomes available to more members of the group. Many carnivores are able to obtain prey of a size that would not be possible if they hunted singly.

(iii) *Reproductive efficiency.* Within the group, help is frequently provided by several adults, male and female, in the rearing of the young. Mutual suckling may occur. Opportunities for breeding success of females are high where breeding aggregations occur. In hierarchial societies it is typically the strongest males that are the most effective breeders. Societies provide an excellent opportunity for kin selection. In some species rearing of large numbers of young is prevented.

(iv) *Optimal use of resources.* General-purpose territories which are retained for long periods probably ensure an adequate provision of resources (food, cover, mating sites, resting places etc.) for the occupants.

(v) *Physiological adaptations.* The clustering of some bats is a mechanism for providing increased insulation and maintaining ambient temperature.

There are certain costs and problems to being social. In general, small defenceless mammals such as rodents and bats are vulnerable to predators, particularly when they assemble in large groups. Thus only the biggest bats occur in large colonies in exposed situations while the smaller colonial species occur in protective roosts. In many territorial systems, there are frequently males (and occasionally females) that fail to obtain territorial occupancy and become surplus. However, these animals may ultimately displace territory holders in which case they are providing a reserve breeding stock. Alternatively, they may disperse and establish themselves in uncolonized areas. Thus what may appear to be (and in some instances actually is) a wastage within the system can be a balancing mechanism between resource availability and the maintenance of an optimum population density.

CHAPTER FIVE

POPULATIONS

The study of populations is particularly important, as it facilitates the description of many ecological processes in quantitative terms. Furthermore, the individual mammal is a recognizable and easily defined unit. By studying populations it is possible to record changes in densities and to quantify the effects of different factors on its numbers. Obviously, densities can vary enormously—for example, in Serengeti, there is one cheetah (*Acinonyx jubatus*) to every 8500 ha (Schaller, 1970); this is in sharp contrast to the 1200+ voles (*Microtus arvalis*) per ha recorded from grassland in Poland (Gromadzki and Trojan, 1971). But it is not the magnitude of densities that is important; it is the reasons for prevailing densities in a locality that require explanation. The study of populations goes a long way to providing this information.

5.1 Survivorship, life tables and age distribution

Mammals are long-lived compared to many insects and other small animals. However, within the group as a whole there is a considerable range in lifespan, from short-lived species such as many rats and mice to long-lived species such as elephants and man. Length of life can be expressed in a survivorship curve where the ages at death of a number of animals (usually corrected to 1000) born at one time are recorded. Examples of nine species broadly representative of the order are illustrated in Figs. 5.1 and 5.2. There is generally a high juvenile mortality. Of all these mammals, the Dall sheep (*Ovis*) has the lowest mortality of young animals, 30% lost in the first year. Thereafter, survival can for a time improve, e.g. hippopotamus, buffalo, or it can continue to decline fairly steadily, e.g. wild

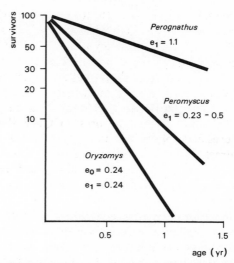

Figure 5.1 High, moderate and low survivorship curves in three small mammals. Life expectancy in years is given at birth (e_0) and one month (e_1) of age (modified from French *et al.*, 1975).

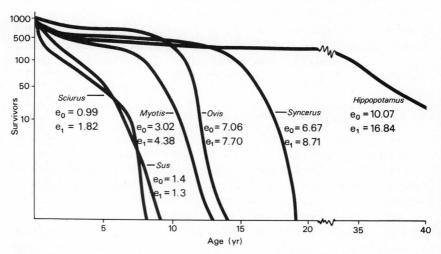

Figure 5.2 Survivorship curves of six species of mammal; grey squirrel (*Sciurus*) (from Barkalow *et al.*, 1970), European wild boar (*Sus*) (from Jezierski, 1977), greater mouse-eared bat (*Myotis*) (from Gaister, 1979), dall sheep (*Ovis*) (from Deevey, 1947), African buffalo (*Syncerus*) (from Sinclair, 1977) and hippopotamus (*Hippopotamus*) (from Laws, 1968). Life expectancy, in years, is given at birth (e_0) and one year (e_1) of age.

boar (*Sus*) and rice rate (*Oryzomys*). In the former category, life expectancy at birth (e_0) is appreciably lower than when the animals are one year old (e_1), whereas in the latter, life expectancy is no greater at one year in *Sus* and one month in *Oryzomys* than it is at birth. There is usually a higher life expectancy with larger size; the bats are an exception—these small mammals often live up to 15 years. In most species survivorship is similar in males and females. A noteworthy exception is the dasyurid marsupial *Antechinus stuartii* (Wood, 1970) of the forests of Queensland. Here, young animals of both sexes first appear in January. By October, all the males are dead, but in contrast, many females survive into their second and third years. The adult population therefore comprises females only from October to June—mating takes place in July and August and soon after this sexual activity the males die.

The elaboration of survivorship data into a life table and the addition of age-specific reproduction (m_x) i.e. the number of progeny produced by each animal within a particular age band, provide further information on demography and population potential. Comparison of a short-lived (*Oryzomys*, Table 5.1) and a long-lived (*Syncerus*, Table 5.2) species highlights the rate at which these two populations replace themselves. In the rice rat, the generation time is 0.41 years, and in buffalo 8.27. The net

Table 5.1 Life and fecundity table for females of the rice rat (*Oryzomys capito*) in Panama (after Fleming, 1971)

Age (months) x	Survivors at start of month l_x	No. of ♀♀ produced by each ♀ m_x	$l_x m_x$	Age specific mortality rate q_x	Life expectancy (months) e_x
0	100	0.00	0.0	28	2.9
1	72	0.00	0.0	29	2.9
2	51	0.00	0.0	29	2.8
3	36	0.99	35.6	28	2.8
4	26	0.99	25.7	27	2.7
5	19	0.99	18.8	32	2.5
·6	13	0.99	12.9	31	2.4
7	9	0.99	8.9	22	2.2
8	7	0.99	6.9	29	1.7
9	5	0.99	5.0	40	1.1
10	3	0.99	3.0	100	0.5

$$\Sigma l_m m_x = 116.8$$

$$R_0 = \frac{116.8}{100} = 1.168$$

Table 5.2 Life and fecundity table for female buffalo (*Syncerus caffer*) in Serengeti (after Sinclair, 1977)

Age (years) x	Survivors at start of year l_x	Number of ♀♀ produced by each ♀ m_x	$l_x m_x$	Age specific mortality rate q_x	Life expectancy (years) e_x
0	1000	0.00	0.0	330	6.67
1	670	0.00	0.0	140	8.71
2	576	0.00	0.0	19	9.05
3	565	0.06	133.9	21	8.22
4	553	0.14	77.4	38	7.32
5	532	0.41	218.1	30	6.66
6	516	0.41	211.6	56	5.85
7	487	0.41	199.7	41	5.17
8	467	0.41	191.5	75	4.37
9	432	0.41	177.1	120	3.68
10	380	0.41	155.8	116	3.12
11	336	0.33	110.9	217	2.46
12	263	0.33	86.8	255	2.00
13	196	0.33	64.7	286	1.57
14	140	0.33	46.2	350	0.99
15	91	0.33	30.0	473	1.35
16	48	0.33	16.0	542	1.10
17	22	0.33	7.3	682	0.82
18	7	0.33	2.3	1000	0.50

$$\Sigma l_x m_x = 1629.3$$

$$R_0 = \frac{1629}{1000} = 1.629$$

reproductive rate (R_0), or the mean number of progeny produced by each female during her lifetime within the cohort, is 1.168 for the rat and 1.629 for the buffalo. The rat inhabits the tropical forests of Panama, where a female living out her maximum life span can produce 6.1 litters averaging 3.9 animals in a year. This represents a total annual productivity of 23.8 animals. In contrast, the mature female buffalo produces one calf at an average interval of 455 days. Expressed differently, the rat would have approximately 20 generations in the time the buffalo had one. If throughout that time the life table characters remained unaltered, the rat's population would increase to over 22 000 and the buffalo's to 1629.

Two generalizations can be made concerning population increase in mammals. Firstly, reproductive capacity (i.e. number of young produced/ animal/year) increases as survival rate decreases. This has been demonstrated for several groups of small mammals (Fig. 5.3). Conversely,

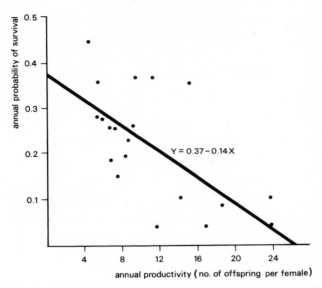

Figure 5.3 The relationship between annual productivity and survivorship in 20 species of tropical rodent (modified from Fleming, 1975).

among the long-lived ungulates reproductive rates are low. Secondly, large mammals frequently have low reproductive rates and small mammals high ones. Thus most artiodactyls are large, and have small litters and long gestation periods, while many rodents produce litters more frequently. There are numerous exceptions to this generalization. For example, the mouse-eared bat (*Myotis*) is small and has a low reproductive rate (one offspring per year) and many small tropical animals (Zapodidae, Sciuridae, fossorial and heteromyid types) have reproductive capacities of four to eight (see Fig. 5.3) which are not dissimilar to the appreciably larger wolf (*Canis*), red fox (*Vulpes*) and rabbit (*Oryctolagus*).

Attention has already been drawn to reproductive flexibility in response to changing environmental conditions (Chapter 3). There can also be changes in mortality patterns. This means that in natural populations demographic parameters such as life table structures, survivorship and age-specific reproduction are not fixed and are subject to change. This is apparent from cohort survival of the wood-mouse (*Apodemus*) in woodland in the west of England (Fig. 5.4), where, for example, the February 1975 cohort (i.e. those animals appearing at this time) had a higher survival rate than those of February 1974 and March 1975. These changes

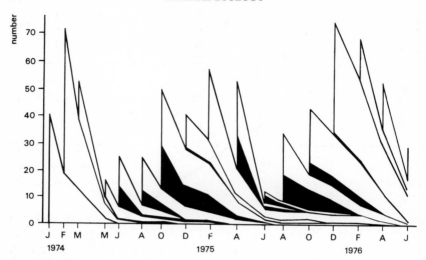

Figure 5.4 Monthly cohort survival of the wood-mouse *Apodemus sylvaticus* in Gloucestershire, England. Shading indicates individuals captured initially as juveniles (modified from Montgomery, 1980).

also mean that age distributions differ at different times. In October 1975 the proportion of animals in the youngest age group is substantially higher than in the same month in 1976.

Age distribution will inevitably change through the year where there is seasonal breeding. The differences are most marked in small mammals where life expectancy is short. In the punctated grass-mouse (*Lemniscomys striatus*), inhabiting the grasslands of western Uganda, there are two discrete breeding seasons coinciding with the wet seasons in March–April and September–November. In June and December (Fig. 5.5) substantial numbers of young animals are caught, and as these animals age they gradually move into older age classes, thereby altering the age distribution. Here the youngest animals in June are those producing the cohort born in November and December.

5.2 Fluctuations

The effects of differing and changing demographic parameters manifest themselves in the fluctuations found in natural populations. These fluctuations can be viewed as changes either within a twelve-month period or over a span of several years. The former reveals seasonal changes while the latter illustrates differences in peaks and troughs from one year to

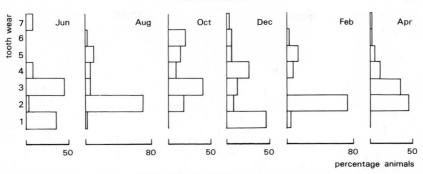

Figure 5.5 Age structure changes in populations of the punctated grass-mouse (*Lemniscomys striatus*) in Uganda in alternate months of the year (from Neal, 1977).

another. Mammals display a number of patterns in respect of both the short and long time periods as the following examples show. A 29-year survey of wood-mice (*Apodemus*) and bank voles (*Clethrionomys*) in Wytham Wood, Oxford (Fig. 5.6) estimated abundance in early summer (May–June) when numbers would be expected to be at a minimum, and winter (November–December) when they would be at a maximum. (This was not always the case, as the appearance in some years of a large acorn or beech mast crop resulted in breeding through the winter and a boosted spring population). The infrequency of the readings masks the more detailed population changes taking place (e.g. Fig. 5.4) but this is nevertheless an extremely useful record of the magnitude of long term changes.

A census of the red deer (*Cervus elaphus*) on the island of Rhum was carried out from 1957 to 1967 (Fig. 5.7). These animals were culled between July and January, when (in all but the first and last years) between 224 and 282 animals were removed. Comparison of the small rodents with the deer highlights the differences between a widely fluctuating population and a relatively stable one. In the former, there can be large differences between peaks and troughs, the population can rapidly change its density upwards or downwards, the peaks are of variable magnitude and may occur in either summer or winter, and the population can display periods of relative stability, e.g. *Clethrionomys* from 1967–69. In contrast, the deer numbers fluctuate regularly within narrow limits. During the 11 years of estimates, the minimum population recorded was never less than 0.7 of the maximum. The peaks and troughs invariably occurred at the same time of year while depletion and recovery were relatively slow. These examples illustrate the differences between unstable and stable populations.

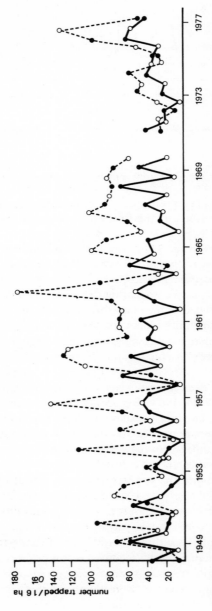

Figure 5.6 Population changes in Wytham Wood, Oxford, in wood-mice (*Apodemus sylvaticus*) (solid line) and bank vole (*Clethrionomys glareolus*) (broken line) from 1949 to 1977. Open circles, summer, (May–June) estimates and solid circles, winter (November–December) estimates (modified from Southern, 1979b).

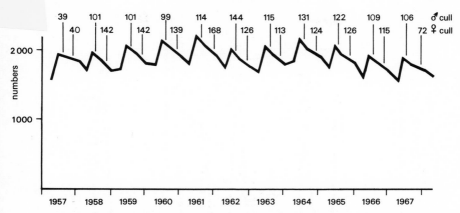

Figure 5.7 Population dynamics of red deer (*Cervus elaphus*) on Rhum, 1957–1968. Stags were culled July to October and hinds October to January (data supplied by Lowe).

It may be argued that culling the red deer resulted in a manipulated stabilization of the population. On the other hand this fairly steady rate of removal could be regarded as similar to the action of a predator in a situation where there is little natural predation. Furthermore, it would be difficult to devise a culling strategy that would stabilize the unstable rodent population. The deer exemplifies a population providing a regular crop or sustained yield. In these cases, culling strategies can be modified from year to year to balance the effects of other mortality factors.

A further type of fluctuation is the cycle. Here, populations appear at peak numbers with remarkable regularity. Cycles occur in several species of birds and mammals living in northern latitudes. As some are important fur bearers, records of relative abundance have been kept by the companies dealing in the skins for well over 150 years. One such species is the lynx (*Lynx canadensis*, Fig. 5.8) which has peaks and troughs approximately every ten years. These cycles are synchronized across the conifer zone of North America, and follow the cycles of the principal prey species, the varying hare (*Lepus americanus*). Apart from their regularity, the most notable features of these cycles are the rapidity of climbs and crashes and the differences between peaks and between peaks and troughs. Interestingly, the hare population is cyclic in some places where the lynx is absent, e.g. Anticosti Island, Canada, and does not cycle on the southern edge of its range. There is also a shorter phase cycle of 3–4 years duration. This is typically found in lemmings (*Lemmus*) and voles (*Microtus*) (section 5.8).

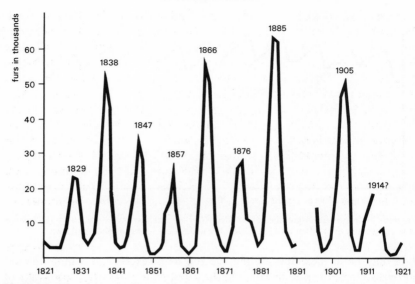

Figure 5.8   The nine-to-ten year cycle in lynx (*Lynx*) in Canada based on fur returns to Hudson's Bay Company (modified from Elton and Nicholson, 1942).

5.3 Mortality

While age at death determines the shape of the survivorship curve, it takes no account of the cause of death. Mammals die in many ways, although in some cases what finally kills an animal may be a mere technicality. For example, a sick and ailing animal may be finally killed by a predator but had it not gone in this way it would soon have died of disease. The main mortality factors are summarized below, together with comments on their effects on population numbers.

Predation. It is difficult to make generalizations about the regulatory role of predators. In unstable populations, predators frequently have difficulty increasing their numbers to keep pace with food supply and it is when the populations fall to low levels that predators have their greatest impact. It is revealing that the tawny owl (*Strix aluco*) population in Wytham Wood, Oxford remained remarkably stable at 20 to 33 pairs throughout the considerable fluctuations of the small rodents (Fig. 5.6) on which it fed.

An example of the differing effects a predator can have on a prey population is found in the spotted hyaena (*Crocuta crocuta*) in Tanzania (Kruuk, 1972). In Serengeti, hyaena follow the migrating herds of wildebeest, zebra and gazelle. During this time adults may have to leave their cubs for several days at a time to go in search of food. As a result, some cubs die of starvation or abandonment. This mortality results in a depression of the hyaena population and prevents it exploiting the herbivores to full advantage. This is in contrast to the situation at Ngorongoro, where the ungulates are not migratory, and where as a result there is higher survival among the hyaena cubs. Here annual recruitment (11%) into the wildebeest population closely matches the take by the predator, so that here it is more likely to assume a regulatory role. Among some large mammals, the delicate relationship between predator and prey is extremely important in maintaining population stability in both species, e.g. the wolf–moose interaction of Isle Royale (section 5.6). Under these circumstances there are often highly evolved anti-predator or alternative ecological devices that ensure only a modest success rate for the predator.

Floods and fire. Large mammals can often escape floods and fire whose advance is often quite slow. It is only when they become isolated on islands when the waters are still rising and fires advancing that they are likely to succumb. For small mammals the picture is different. Many cannot escape and must die, as in the periodic flooding of the Kafue Flats in Zambia (Sheppe and Osborne, 1971). It has, however, been noted in African grasslands that there is considerable post-burn survival of small rodents (Neal, 1970). Here, as the fire sweeps across the ground, many rodents seek refuge in holes in which they survive. They subsequently surface on a charred and exposed landscape where they often become the ready victims of predatory birds and mammals.

Undernutrition is the shortage of one or more of the main constituents of diet, which, at its most severe, results in death. There are numerous examples of this. In temperate regions in a severe winter, normally accessible food is no longer available and deaths can result, e.g. amongst deer and rabbits in Europe. The analogous situation in the tropics is when there is an occasional failure of the wet season resulting in drought and a poor production of vegetation. This occurred in East Africa in 1961 and 1971 when appreciable numbers of animals died (Corfield, 1973; Delany and Happold, 1979). This may represent an extreme of what is probably commonly taking place in seasonal habitats. Here, in the adverse season, insufficient quality food is available and as a result use is made of reserves built up in the favourable season. African buffalo, elephant and wildebeest all experience food shortages during the dry season. For some animals there is a delicate balance between survival and death, with a small number failing to live through this time. Undernutrition can result in debilitation and increase vulnerability to death from other causes.

Diseases and parasites. Diseases can sweep through populations and dramatically reduce their magnitude. Well-known examples include the myxomatosis virus in rabbits in Europe and Australia, and the great rinderpest epidemic that entered tropical Africa about 1889 and reached the Cape by 1896 (Ford, 1971). This disease killed buffalo, eland, bushbuck (*Tragelaphus*) and giraffe on a massive scale. A few animals escape diseases such as these because of immunity or other factors. There is frequently a recovery in numbers following a severe outbreak but, as with these two examples, even an exotic disease remains as a continuing mortality factor. The plague bacterium (*Yersinia pestis*) is known to kill large numbers of rodents. Of 34 gerbil (*Tatera brantsi*) colonies examined in 1940 in the Orange Free State, South Africa, only three survived a plague epidemic (Davis, 1953). There are many other examples of rodent-transmitted human disease e.g. Lassa fever, although little information is available on their effects on mammal populations. Further examples of diseases in mammals are bovine tuberculosis in badgers and rabies in foxes.

Intraspecific mortality. Deaths can result from parental killing of young (as in lions and wild dog), in defence of territory (e.g. mice, wolves), and through the driving out of animals, particularly young, from social groups (Chapter 4). These surpluses may find themselves vulnerable in hostile environments and subsequently die of other causes.

Physiological fatigue is found in the male of the small Australian marsupial *Antechinus stuartii*. Immediately following breeding there is a physiological decline and accelerated senility. There is no evidence that this is related to sexual activity and fighting, as isolated males display the same deterioration (Woolley, 1966).

5.4 Natality

The factors influencing numbers of young produced have been considered in Chapter 2. For present purposes they are categorized and summarized in Fig. 5.9. The situation is more complex than illustrated, as several examples of feedback mechanisms are known to occur which have a profound effect on numbers of young produced. When this is high, popula-

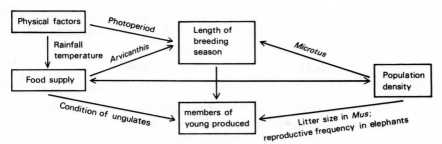

Figure 5.9 Interactions of physical and ecological factors that influence population density. Examples of interactions are indicated and discussed in text.

tion density will increase and food supply will decline. At high population densities, elephants for example produce few young, the population gradually declines, but subsequently increases again through increased natality. Thus while natality is under the control directly or indirectly of abiotic influences, it is also highly adaptable for density regulation.

5.5 Dispersal

A dispersing animal is one that makes a permanent move from its home area. This can be initiated by forced eviction as in many social species, or by voluntary movement. Among the latter are many mammals, e.g. *Oryctolagus*, *Rattus* and several primates, that are known to make occasional exploratory forays from their home range. With time these newly-explored areas can become permanently occupied. Some young bats, e.g. *Tadarida brasiliensis* and *Miniopterus schreibersi*, disperse from the maternal colony soon after weaning and frequently subsequently occupy an alternative roost. This interchange is similar to the movement of males from one social unit to another, as is found in mammals as different as gorillas (*Gorilla*) and vicunas (*Vicugna*). Finally, dispersal can occur if the habitat is changing unfavourably. In times of drought, the muskrat (*Ondatra zibethicus*) first moves to a base at the periphery of its familiar area and from there makes sorties into new areas. Dispersal movements have been comprehensively reviewed by Baker (1978). It has been suggested that when dispersal takes place at presaturation densities, as with some rodents, populations may not attain the carrying capacity of the habitat (Lidicker, 1975).

As dispersal assumes a number of forms, so does its role in population processes. Eviction probably serves to maintain evictors at optimal density. Other important functions include genetical interchange between demes or larger groups, and the potential exploitation of unoccupied and temporarily suitable habitats.

5.6 Growth and stabilization of populations of large mammals

The growth of populations of large mammals can be considered when animals enter a new, previously uninhabited but favourable habitat, or when they are re-established in an area previously inhabited but where (owing to some extraneous or unnatural circumstance) they have either been eliminated or reduced to very low numbers. Within the former are several examples of artificial and natural introductions. Under both conditions the population could be expected to follow a logistic growth curve. The predicted course of events has been modelled (Fig. 5.10). Herbivores initially overshoot the optimal resource level and subsequently

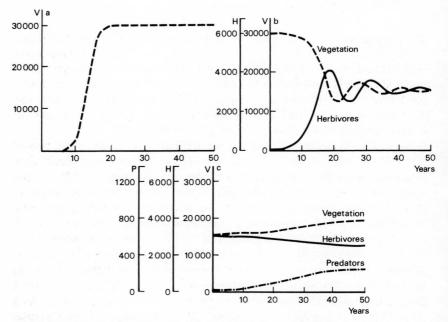

Figure 5.10   Models of (a) ungrazed plant population growth (V), (b) a population of plants (V), and the herbivores (H) feeding on it and (c) a population of predators (P) and herbivores on which it feeds and the plants sustaining the herbivores. Figures are expressed in biomass (modified from Caughley, 1977).

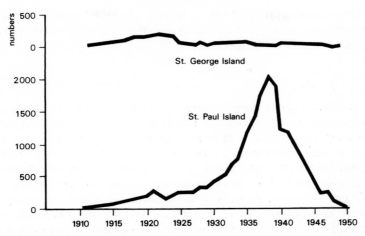

Figure 5.11 Population changes in reindeer (*Rangifer caribou*) on the Pribilof Islands (modified from Scheffer, 1951).

show a reduction in numbers. A continuing and declining oscillation results in stabilization after about 40 years. The introduction of carnivores into a stable herbivore-vegetation situation results in a logistic growth of the predators accompanied by a small decline in the herbivore level and a small increase in the vegetation. A major shortcoming of models, however, is that they oversimplify and reduce complex natural situations to unrealistically few variables.

Two examples of introduced species are reindeer (*Rangifer*) on two of the Pribilof Islands off the coast of Alaska in 1911, and mule deer (*Odocoileus*) on the 485 ha George Reserve, Michigan, in 1928. In the tundra of St. Paul Island, the reindeer population initially grew slowly (Fig. 5.11) and then grew rapidly, considerably exceeding the estimated 800 that the island could comfortably support. In contrast, on St. George, the other Pribilof Island, the reindeer population never reached the assumed supportable level. In George Reserve, Michigan, the deer introduced increased their numbers rapidly so that within seven years they exceeded 200 (O'Roke and Hamerstrom, 1948). By this time habitat deterioration was evident, with no sign of deer numbers declining. As a result they were regularly reduced by culling. In none of these examples was a stable population attained naturally that was in balance with the food resources. Furthermore, there were considerable differences in the timing and extent of realization of maximum biotic potential.

A natural introduction occurred on Isle Royale in Lake Superior when a small number of moose (*Alces*) reached the island from the adjacent mainland in about 1908 (Mech, 1966). These herbivores soon established themselves, finding ample, unexploited food. Numbers increased, so that by the late 1920s there were several thousand present. Subsequent sporadic observations during the next 20 years indicate sequences of gross over-consumption, habitat deterioration, dramatic population declines and modest recoveries which never attained the earlier maximum. Here was another example of a population failing to achieve a balance along the lines of the proposed model. Wolves (*Canis*) first reached the island in 1948, and their numbers gradually increased until about 1960 when they stabilized with the moose population. It is possible that through the 1950s the wolf population followed a logistic curve; unfortunately no records exist of its numbers at this time. But it does appear that the herbivore came into equilibrium in accordance with the model when the predator was present.

The European bison (*Bison*) occurred in large numbers in the forests of eastern Europe. Records from Bialowieza Forest in eastern Poland indicate a population of 650 to 750 animals from the turn of the century to the end of the First World War, when they were eliminated (Krasinski, 1978). In 1952, bison were reintroduced, and with the help of sup-plementary winter feed their numbers steadily increased (Fig. 5.12) to 253 in 1973. Over this period there has been a gradual decline in reproductive success. From 1958 to 1969 the mean annual percentage of

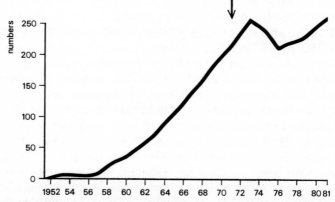

Figure 5.12 Population changes in bison (*Bison bonasus*) in Bialowieza Forest, Poland. Arrow indicates when culling commenced (data supplied by Krasinski and Pucek).

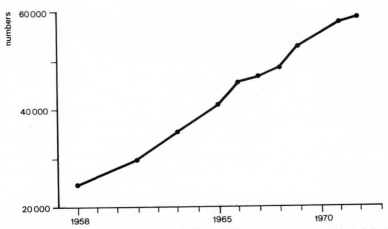

Figure 5.13 Increase in number of buffalo (*Syncerus caffer*) in Serengeti, Tanzania following the rinderpest outbreak of the 1950s (modified from Sinclair, 1977).

calves born to adult cows was 69 and for 1970 to 1973 this figure fell to 49. Even so, the population was increasing rapidly and the decision was taken in 1971 to initiate culling to maintain the herd at about 230. Seventeen animals were removed in the first two years, and in 1975 and 1976, 44 and 57 respectively; for the remainder the annual removal was between 15 and 25.

Periodic outbreaks of the exotic virus of rinderpest result in rapid, high mortality of certain species of African ungulates. The buffalo (*Syncerus*) of Serengeti were the victims of rinderpest in the 1950s. Thereafter, there was a recovery which was levelling off by the early 1970s (Fig. 5.13). In this locality, where there was a full spectrum of other mammal species, the population displayed logistic growth.

"Carrying capacity" is a term frequently used by managers of large herbivore populations. It has been defined (Petrides and Swank, 1965) as the maximum numbers or biomass of animals which can be supported within a given area for an indefinite period. Thus, under natural conditions, this refers to the supportable population at the most adverse time of year in the most adverse year. When applied to many large mammals whose habitats and populations fluctuate within modest limits this is a useful term. It is of less value for small mammals whose populations frequently fluctuate between wide limits and for those large mammals living in situations where occasionally numbers are dramatically reduced. The usage of the term does not always comply with the rigidity of the above definition. It is sometimes used synonymously with K, the upper asymptote of the logistic growth curve, or alternatively for the number of animals a habitat is known to support for long, but not necessarily indefinite, periods of time.

5.7 Population regulation

The regulation of mammal populations can be briefly illustrated by reference to two well-studied species of differing size and habitats. They are the wood-mouse (*Apodemus*), a small rodent of Palaearctic woodlands, and the African buffalo (*Syncerus*), a large ungulate of tropical savannas and forests.

Detailed analysis of the causes of the wood-mouse fluctuations in Wytham Wood (Fig. 5.6) indicates the involvement of several factors. It was noted that populations were usually at about the same level in mid-winter and that subsequent events during the rest of the year dictate the course the population follows. If there is a plentiful winter food supply, breeding continues and the winter drop in the population is small (Fig. 5.14). If, in contrast, there is a poor supply of winter food, the population drops dramatically. The lower the level then reached by the spring

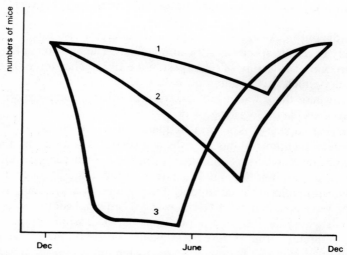

Figure 5.14   Population changes in wood mice (*Apodemus sylvaticus*) in Wytham Wood, Oxford, in years when winter food is (1) abundant, (2) moderate, (3) scarce (modified from Watts, 1969).

population the sooner it commences its recovery. However, the timing of the recovery is remarkably late. In most years this is not until the latter part of the summer or early autumn even though breeding will have started in early spring. This apparent anomaly has been attributed to the behaviour of the overwintering adult males, although the precise mechanism they use is uncertain. They could be involved in dispersing young animals or in some way preventing their survival. Their influence is nevertheless one of suppressing the population, and the larger the population the longer they fulfil this role. Suddenly this pressure is released and the population grows rapidly. Again the causes are not known with certainty, but speculation must revolve around the disappearance of these dominant males or a decline in their agonistic behaviour. Summarizing, the important regulatory factors in this animal are food supply in winter, the self-regulatory role of adult males in spring and summer, and the sudden release of this effect permitting the population to return to a steady winter level.

Natality and mortality data were examined in the buffalo from 1965 to 1973, and annual changes in female fertility were also recorded. Analysis of all these data showed juvenile mortality to be the major cause of population fluctuation. This is density-independent and results from several causes whose intensity varies randomly from year to year. In contrast, adult mortality is density-dependent, with proportionately more deaths in higher populations. Although this is a less significant mortality factor in that it has less influence on total numbers than does juvenile mortality, it operates so as to compensate for the effects of juvenile mortality and thereby dampen fluctuations. Changes in female fertility have little effect on regulating the population. Mortality is caused by several factors (Fig. 5.15), with undernutrition identified as of major importance through its direct and indirect effects. The quality and quantity of food changes seasonally, and as a result of inadequacy at certain times of the year buffalo are then more vulnerable to predators and disease. There are also several positive and negative feed-back mechanisms. For example, undernutrition can result in behavioural modifications that lead to more effective exploitation of alternative food resources, which thereby gives the original source greater opportunity for recovery.

5.8 Cyclic species

Cyclic changes in the populations of small rodents have attracted the most detailed investigation. This is probably not surprising as the relatively

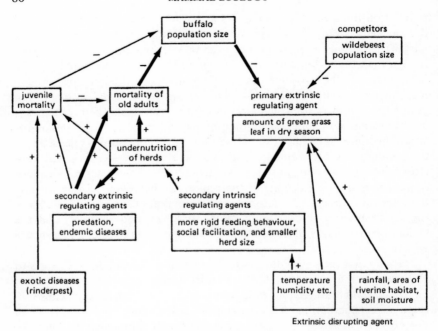

Figure 5.15 Model of effects of various factors on the regulation of buffalo (*Syncerus*) populations. The main pathways are shown by heavy lines. Minus signs indicate that a higher intensity has negative results; plus signs indicate positive results (modified from Sinclair, 1977).

short interval of 3–4 years between peaks makes them more amenable subjects of study than those animals (e.g. showshoe hare) having a longer interval (9–10 years). The rodents examined in greatest detail belong to the genera *Microtus* and *Lemmus*, from the North American sub-arctic tundra to the more southerly temperate grasslands. As will be seen, there is evidence that the population processes are not the same in these two habitats.

The population cycle of *Microtus pennsylvanicus* in grassland displays a number of characteristic features (Krebs and Myers, 1974). During the increase phase, mortality is low, there is considerable dispersal, the breeding season is relatively long and the adults attain a larger size than at other times. Dispersal continues through the peak phase. When dispersal was experimentally prevented by large field enclosures, considerable overgrazing of the vegetation ensued because the population attained densities greatly in excess of the peak of unenclosed populations. At high

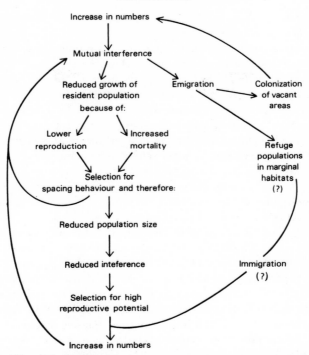

Figure 5.16 Chitty-Krebs hypothesis explaining cyclical population fluctuations in rodents (modified from Krebs *et al.*, 1973).

densities, the breeding season was shorter and the onset of sexual maturity delayed. The population decline can be either rapid or gradual, lasting for up to two years. In the event of the latter there is the implication that resources are not in dramatically short supply after the peak. It is important to appreciate that throughout the cycle animals are constantly replacing themselves, so that different animals are present at different phases of the cycle. This has led to the proposal, supported by limited evidence, that the genetic composition of individuals differs from one phase of the cycle to another. Thus, as the population attains high levels, the aggressive genotypes are favoured, which manifest themselves as the cause of dispersal. As the population declines, the less aggressive genotypes are favoured. This analysis of *Microtus* populations has led to the Chitty–Krebs hypothesis (Fig. 5.16) which stresses the importance of self-regulatory factors in these cycles.

This state of affairs differs from what happens on the tundra (Batzli,

1975), where at times of population peaks, North American brown lemmings (*Lemmus trimucronatus*) do immense damage to the vegetation through consumption and burrowing. These animals show rapid climbs and crashes in populations. The latter could be the result of the rapid depletion of resources. In the European lemming (*Lemmus lemmus*) there can be massive dispersal at this time from tundra to coniferous forest. The differences between *Lemmus* and *Microtus* may be related to habitat conditions. For the latter, the provision of grassy cover may be important, and the social behaviour could be as easily directed at ensuring this as at providing an adequate food source. Where there is burning and grazing of *Microtus* habitats, and the populations are generally low, there is no cycling of the population. The apparent lack of uniformity in population processes may then be the result of, or considerably influenced by, the type of habitat being occupied.

Predators, both birds of prey and ground-dwelling carnivores, may exercise an influence on population cycles. However, when the cycle is going through an increase phase, the reproduction of the rodents outstrips that of the predators, and although predator numbers do increase, their overall effect at this time is minimal. It is during the decline phase and the subsequent period of low density that they probably have their greatest effect. As they chase less and less food, their relative efficiency increases, and they then suppress the rodent population to a low level for a time (Pearson, 1966, 1971; Fitzgerald, 1977). But when very few voles are available, predator numbers decline, and the rodents are then in a position to multiply freely.

CHAPTER SIX

NICHE, SPECIES DIVERSITY AND INTERACTION

The land masses of the world are covered by a wide range of vegetation types whose composition is determined largely by climate and soil.

It is possible to broadly classify much of this vegetation into a small number of major categories or plant communities (Table 6.1). Several

Table 6.1   Some important plant communities and their distributions (after Vaughan, 1978)

Community	Distribution
Tropical rain forest	South and Central America; Africa; S. E. Asia; East Indies; N. E. Australia
Tropical deciduous forest	Mexico; Central and South America; Africa; S. E. Asia
Temperate rain forest	Pacific Coast from northern California to northern Washington; parts of Australia, New Zealand and Chile
Temperate deciduous forest	Eastern USA; parts of Europe and Eastern Asia
Subarctic-subalpine coniferous forest	Northern North America; Eurasia; high mountains of Europe and North America
Thorn scrub forest	Parts of Mexico; Central and South America; Africa; S. E. Asia
Temperate woodlands	Parts of western and southwestern USA and Mexico; Mediterranean area; parts of Southern Hemisphere
Temperate shrublands	California; Mediterranean area; South Africa; parts of Chile; West and South Australia
Savanna (tropical grasslands)	Parts of Africa, Australia, southern Asia, South America
Temperate grasslands	Plains of North America; steppes of Eurasia; parts of Africa and South America
Arctic and alpine	Tundras north of treeline in North America and Eurasia; some areas in Southern Hemisphere
Deserts	On all continents; widespread in North America, North Africa and Australia

occur in more than one continent, and, while it is relatively easy to recognize a community-type (e.g. grassland), it is probable that the species of plants comprising that community will differ from continent to continent. Furthermore, plant communities frequently merge imperceptibly into each other as conditions gradually change.

The number of mammal species supported by a community depends on several factors. These include the level of primary production (which accounts for the quantity of food produced), seasonal availability of resources, floral heterogeneity, diversity of plant structure, nature of the substratum and previous history of the community. Tropical situations generally have higher production than their temperate equivalents, and forests are more productive than grasslands. Tundra and alpine situations have low production and deserts the lowest of all. Forests add a spatial dimension with their vertical stratification—in tropical rain forests, up to five distinct layers are recognized. Floral diversity affords greater opportunity for the evolution of monophagous feeders, while seasonality or aseasonality influences the number of adaptive types that can be supported. For example, nectar-feeding bats and fruit-eating primates and bats can survive only in situations where there is a constant and plentiful supply of these foodstuffs. These considerations point to a richness of species in aseasonal tropical rainforests where there is high production, an abundance of plant species and considerable structural diversity. This is supported by the large number of species from the forests of West Africa, Central America and South-East Asia (Fig. 6.1). In all three forest regions there are numerous species of frugivorous and insectivorous bats, and arboreal mammals. Comparing this situation with the tropical African grassland where there are only scattered trees, there is, as would be expected, an appreciable drop in arboreal species and (with most primary production taking place at ground level) a large number of terrestrial species. The temperate deciduous forests of Michigan and England contain less rich faunas with, in the latter, few arboreal species. Finally, both the subarctic tundra and the sand plains of the Sahara support a small number of ground-dwelling species.

These considerations of species diversity lead to the concept of ecological niche. The niche can be regarded as the sum of all ecological requirements of a species. This includes food, space and shelter as well as the acceptable physical conditions. The fundamental niche has been defined as the n-dimensional hypervolume whose axes describe different environmental variables. The volume occupied by the species is that which permits its indefinite survival. The actual niche occupied by the species in

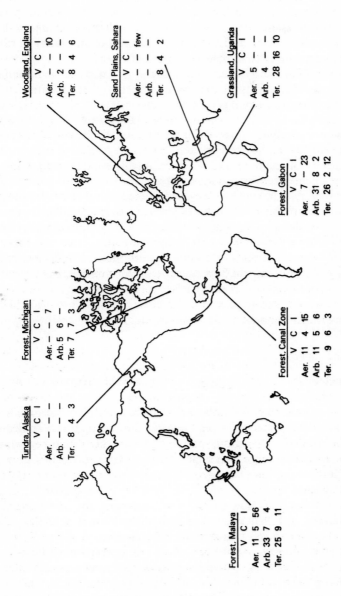

Woodland, England

	V	C	I
Aer.	–	–	10
Arb.	2	–	–
Ter.	8	4	6

Sand Plains, Sahara

	V	C	I
Aer.	–	–	few
Arb.	–	–	–
Ter.	8	4	2

Grassland, Uganda

	V	C	I
Aer.	5	–	–
Arb.	4	–	–
Ter.	28	16	10

Forest, Gabon

	V	C	I
Aer.	7	–	23
Arb.	31	8	2
Ter.	26	2	12

Forest, Michigan

	V	C	I
Aer.	–	–	7
Arb.	5	6	–
Ter.	7	7	3

Tundra, Alaska

	V	C	I
Aer.	–	–	–
Arb.	–	–	–
Ter.	8	4	3

Forest, Canal Zone

	V	C	I
Aer.	11	4	15
Arb.	11	5	6
Ter.	9	6	3

Forest, Malaya

	V	C	I
Aer.	11	5	56
Arb.	33	7	4
Ter.	25	9	11

Figure 6.1 Numbers of species in various world vegetation types. V, vegetarian; C, carnivore; I, insectivore; Aer., aerial, Arb., arboreal, Ter., terrestrial (data largely from Delany and Happold, 1979; Fleming, 1973; Harrison, 1962).

the wild may be smaller than this, and is referred to as the "realized niche". The breadth of the niche may differ for different species. For example, the omnivorous Eurasian wood-mouse (*Apodemus sylvaticus*) has a large geographical range, from Ireland to the Himalayas and from northern Europe into Arabia. This ubiquitous small rodent, while preferring deciduous woodland and forest edge, is found in a wide range of habitats from grasslands and fields in Britain to the sub-alpine habitats of the Middle East. Such an animal is a broad-niche species. This contrasts with the narrow-niche Delany's swamp-mouse (*Delanymys brooksi*) whose habitat and geographical range are extremely restricted. This small granivorous rodent is known only from a few high altitude sedge-swamps on the borders of Uganda and Zaïre. Many species of mammal have overlapping niches sharing a common resource for at least part of their lives. A difficulty confronting the field ecologist involves establishing whether the identifiable field niche of a species is the niche of its choice and adaptation, or whether it is confined within certain ecological limits by the activities of other species. The extent of this competitive exclusion is poorly understood for most species of mammals, although, as will be seen, the partitioning of resources and how they are shared between closely related species inhabitating a common area are better known. The competitive exclusion of a species from part of its range has the result that its realized niche is that much smaller than its fundamental niche.

6.1 Use of food and space

The tropical forests with their complex vegetational structure and floristic variety support a richness of species which permits examination of resource use at its most complex. Analysis of the microdistribution of small rodents (rats, mice, gerbils) in a central African forest (Fig. 6.2), indicates a remarkable spatial separation, particularly at the ground-vegetation interphase. Examination of their diets, when account is taken of size, points to further separation—*Lophuromys* is largely an insect eater, and the three terrestrial species from drier areas are of quite different sizes, ranging from weights of 8 g in *Mus* to 100 g in *Aethomys*. The diet and vertical distribution of seven species of sympatric diurnal squirrels in the Malaysian rain forest shows distributional differences with some species, e.g. black giant squirrel (*Ratufa bicolor*), preferring only one zone of the forest, and others, e.g. horse-tailed squirrel (*Sundasciurus hippurus*), roaming more freely up and down the vegetation (Fig. 6.3). Observations on feeding indicate further ecological separations. The plantain squirrel

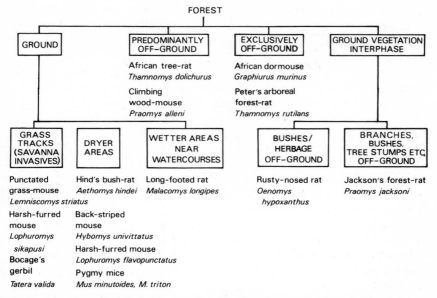

Figure 6.2   The ecological distribution of small rodents in a tropical forest in Uganda (from Delany, 1971).

(*Callosciurus notatus*) consumes a wider range of food (flowers, soft fruit, leaves, bark) than does the horse-tailed squirrel (*S. hippurus*) (hard and soft fruit, leaves). Two squirrels (*R. bicolor* and *C. prevostii*) occupy the same part of the habitat and both largely feed on fruit. There is, however, an appreciable size difference between these two species—the giant squirrel (*Ratufa*) weighs up to 1620 g and Prevost's squirrel (*Callosciurus*) up to 505 g. In these forests squirrel densities are low. This is probably the result of competition with other mammals, e.g. tupaiids, primates, and flying lemurs, and the irregular fruiting patterns of many trees. In addition to these diurnal squirrels, there are also present two species of nocturnal flying squirrels.

Examination of the diets of the four commonest species of rodent inhabiting the *Bouteloua–Buchloe–Carex* short-grass prairie of Colorado (Fig. 6.4) shows both species preferences and seasonal changes. The kangaroo rat (*Dipodomys*) is mainly granivorous throughout the year, although herbs form a significant minor component in early summer. While the deer-mouse (*Peromyscus*) feeds extensively on seed in autumn and winter, for the rest of the year insects comprise an important element

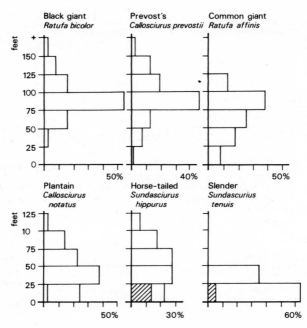

Figure 6.3 Vertical distribution of sightings of diurnal squirrels in tropical forest in Malaya. Shading indicates sightings on ground (modified from Mackinnon, 1978).

in the diet. The grasshopper mouse is an omnivore with insects forming the bulk of the diet throughout the year; seeds are only a minor constituent. The 13-lined ground squirrel (*Spermophilus*) eats plant and animal materials in approximately equal proportions with no obvious priority for any food.

Trophic relationships are still more complex in situations where a range of plant species are all consumed by several mammals and it is the proportions taken by each species that vary. The comparison of wet and dry season feeding habits of four ungulates living together in Rwenzori Park (Fig. 6.5) show, firstly, that in buffalo, kob and warthog, there is a significant change in diet from one season to another. Of further importance are the grass priorities. In the wet season buffalo favours, in order of preference, *Heteropogon*, *Sporobolus* and *Themeda*, for kob the order is *Themeda* and *Heteropogon*, for hippopotamus *Sporobolus* and *Cynodon* and for warthog there is a more even spread with *Eragrostis*, *Cynodon* and *Sporobolus* the most preferred. Thus there is the possibility of

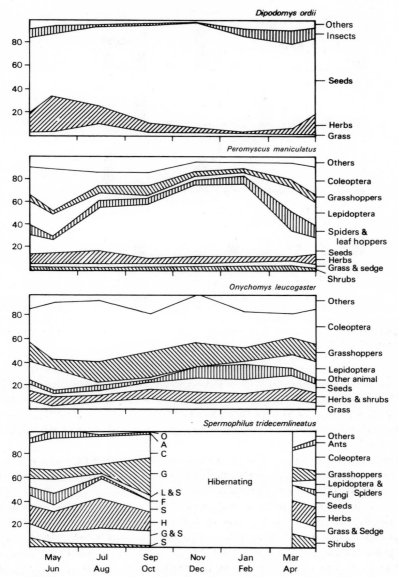

Figure 6.4   The percentage composition by volume of the diet of Ord's kangaroo rat (*Dipodomys ordii*), northern grasshopper mouse (*Onychomys leucogaster*), prairie deer mouse (*Peromyscus maniculatus*) and the thirteen-lined ground squirrel (*Spermophilus tridecemlineatus*) on a short-grass prairie in Colorado (from Flake, 1973).

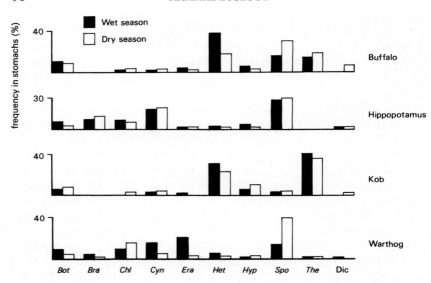

Figure 6.5 Percentage frequency of plant fragments in stomachs of buffalo, hippopotamus, kob and warthog during the wet and dry seasons in Rwenzori Park (Region B), Uganda. *Bot.*, *Bothriochlea*; *Bra.*, *Brachiaria*; *Chl.*, *Chloris*; *Cyn.*, *Cynodon*; *Era.*, *Eragrostis*; *Het.*, *Heteropogon*; *Hyp.*, *Hyparrhenia*; *Spo.*, *Sporobolus*; *The.*, *Themeda*; *Dic.*, dicotyledons (from Field, 1972).

active competition for *Sporobolus* in the dry season. Studies on large herbivores elsewhere in Africa have shown how sympatric species may favour different parts of grasses. In Serengeti, zebra mainly select stem and sheath to a lesser extent, wildebeest sheaths and some leaf, and topi (*Damaliscus*) mostly sheath and some stem. Here again there are overlaps (Gwynne and Bell, 1968).

The introduction and spread of north American mink (*Mustela vison*) in Britain lead to speculation concerning its trophic relations with the indigenous otter (*Lutra*) which occupies the same aquatic habitats. Examination of faeces of both species from lochs and river systems in Scotland, whilst indicating some overlap, generally show fairly distinct feeding habits (Jenkins and Harper, 1980). The main food of otter is fish, notably eels and salmonids and to a lesser extent perch, pike and minnow. Although mink eat small quantities of these items, their major foods are birds (ducks, rails) and mammals (rabbits, field voles) which form only minor components of the otter diet. The mink is apparently more terrestrial in its foraging than the otter.

6.2 Activity

Times of activity are adaptive. For many small, herbivorous mammals crepuscular or nocturnal activity is beneficial as it reduces the likelihood of capture by predators. For some larger mammals, this is less important, and as a result their activity patterns are variable. The hippopotamus retires to wallows during the day and feeds on land at night, in contrast to the elephant, which has its main resting period from 0300 to 0700 hours (Wyatt and Eltringham, 1974). Environmental conditions and physiological demands can also impose limits on activity. In deserts, most mammals are inactive during the hottest times of the day and many of the smaller species retire to secluded domiciles. The high energy expenditure, and consequent high level of food consumption, of temperate shrews such as *Sorex* is probably responsible for a pattern comprising ten equally-spaced active periods during a day (Crowcroft, 1954). Here major activity peaks occur at 2000 and 0400 hours with the lowest peaks between 0700 and 1100 hours. Food availability may be a determinant of activity in mammals such as insectivorous bats which are dependent on the nocturnal appearance of insects. Many temperate mammals alter their behaviour with changing photoperiod—nocturnal rodents such as *Apodemus* have a shorter period of activity in midsummer than midwinter (Brown, 1956).

In situations where several species with similar ecological requirements occur together, temporal separation of activity can provide another dimension in niche separation. For example, in a tropical African grassland supporting four species of omnivorous small rodent, two are abundant and approximately the same size. Of these one (*Lemniscomys*) is crepuscular, and the other (*Praomys*) nocturnal. The two remaining species (*Mus* spp.) are both nocturnal but appreciably smaller; little is known of their habits (Delany, 1964). In the woodlands of Iowa, insectivorous bats have different periods of foraging activity (Fig. 6.6). This is influenced by food specificity and availability. *Eptesicus* feeds on Coleoptera and Hymenoptera, and *Lasionycteris* and *Lasiurus* largely on moths.

When sharing a common resource, the times of maximum use may be species-specific. This has been demonstrated in the use of artificial waterholes in Tsavo Park, Kenya, by 12 species of ungulates. These animals use the waterholes regularly, but with remarkably little temporal overlap of the times of major visits. Rhinoceros, buffalo and elephant are nocturnal visitors, the elephants being most frequent soon after darkness and the other two closer to midnight. Eland and giraffe appear just before and after dawn respectively, and the smaller species, which are more

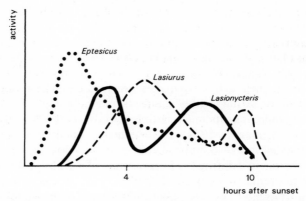

Figure 6.6 Foraging activity of the brown bat (*Eptesicus fuscus*), silver-haired bat (*Lasionycteris noctivagans*) and the hoary bat (*Lasiurus cinereus*) in woodlands in Iowa (modified from Kunz, 1973).

vulnerable to predation, are typically present from mid-morning to mid-afternoon. During this period there is some overlap, notably between warthog, impala, waterbuck (*Kobus*), and zebra in early afternoon. Gazelle (*Gazella*) and hartebeest have their peaks in the morning and oryx (*Oryx*) towards the latter part of the afternoon (Ayeni, 1972).

6.3 Interspecific competition

Competition between two species can be expected when they are using a common resource which is inadequate to meet their combined demand. Among the more important resources are living space, food, breeding sites and shelter. It is possible that competition may be temporary as, for example, at those times when food is in short supply. Competition is frequently difficult to recognize in the field. This is because the overt expression of interspecific competition is seldom witnessed, and the successful exclusion of one species by another is frequently already achieved. Even so, this is not to say that competition is not taking place in situations where species ranges meet and overlaps occur. An indirect approach to the measurement of competitive effects is to examine the ecology of species that occur together in parts of their range and in isolation elsewhere. For this, studies of island populations are frequently helpful. The common shrew (*Sorex araneus*) and pygmy shrew (*Sorex minutus*) are commonly found in the same grassland in mainland Britain, where the former are appreciably more numerous. The common shrew is absent from Ireland and many of the Scottish islands. If competition occurs between these two species then the pygmy shrew, in the absence of the common shrew, could be expected to be more numerous in island habitats similar to those on the mainland. In fact, this has not been found to be the case (Ellenbroek, 1980).

Among the British voles, *Microtus* inhabits grassland and *Clethrionomys* woodland and scrub, with these habitat distinctions quite well recognized on the mainland. Several small off-shore islands support one or other of these species. Where this occurs the resident species occupies both types of habitat. Furthermore, the recent introduction and rapid spread of *Clethrionomys* in Ireland, where *Apodemus* was widespread, illustrates how easily a species may establish itself when a less adapted species is occupying its niche. The *Microtus–Clethrionomys* interaction has been studied in considerable detail in field and laboratory in Canada (Grant, 1962). When one of each species was introduced in a laboratory enclosure equally divided between grass (*Dactylis*) and maple (*Acer*), the two species were mutually dispersive, with each occupying its typical habitat. When only one species was present it occupied the whole area. This work was confirmed in the field, when removal of one species from an enclosure containing grass and woodland resulted in the spread of the remaining species throughout the area. However, further research on natural populations indicated the pattern was not as consistent as this. Instances were recorded of *Microtus* moving into woodlands in November and *Clethrionomys* into grassland in October. On both occasions the two species lived together until the following spring when the invasive moved out. It is suggested that during the winter non-reproductive period, aggression declines in both species, which are mutually tolerant until the onset of spring breeding.

There is an interesting zonation of four species of chipmunks (*Eutamias*) on the eastern slopes of the Sierra Nevada in California (Chappell, 1978). These are *E. alpinus* in the alpine zone (above 3000 m), *E. speciosus* in the lodgepole pine zone (2400–3000 m), *E. amoenus* in the pinon pine-mahogany zone (1900–2400 m) and *E. minimus* in the hot, sagebrush desert (below 1900 m). All four species are generalist feeders consuming similar foods. The reasons for this stratification are complex, although much influenced by competitive interaction and physiological adaptation. There is a linear dominance hierarchy—*E. speciosus* dominates *E. amoenus*, which dominates *E. minimus*. It might then be expected that *E. speciosus* would move into lower altitudes, but this does not happen because the pinon pine-mahogany is more thermally stressful to this species than *E.*

amoenus. Similarly, the hot desert is too inclement for *E. amoenus*. Of these chipmunks, *E. minimus* and *E. amoenus* have the physiological capability to move to higher altitudes, but are prevented from doing so by competitive activity of the species above them. They thus have an appreciable unrealized niche, whereas this is not the case for *E. speciosus*.

Among large mammals competition has already been suggested for grasses among tropical ungulates particularly during the dry season (section 6.1). Dry season competition has been examined in more detail between wildebeest and buffalo in Tanzania (Sinclair, 1974). These species consume much the same parts of the same species of grass. Within the study area there was some spatial separation, with 83 % of the wildebeest preferring open *Themeda* grassland and the remaining 17 % moving into the riverine grassland. The buffalo moved into riverine forest which could support only part of the population and the remainder was forced into the riverine grassland. At this time of year there was a food shortage for the buffalo and no alternative habitat they could move to. Had the wildebeest not been present, it is estimated that a buffalo population 18 % larger could have been supported. In the same area, competition frequently occurs between the large carnivores. Lion and spotted hyaena (*Crocuta*) will each chase the other species off its kill, but hyaenas will more often take the kills of leopard, cheetah and wild dog than lose theirs to them.

6.4 Migration

Migration is a temporo-spatial change of niche having the result that the animal makes use of different situations at different times of the year. It is an annual cyclic event insofar as the same circuit is covered each year. Among migrating mammals are bats, whales (section 2.3) and ungulates.

Migration distances in bats vary considerably. A population of 300 000 little brown bats (*Myotis lucifugus*) hibernates in the Mount Aeolus cave in Vermont (Davis and Hitchcock, 1965). In spring, these animals radiate into New England where they take up summer residence under warm roofs and in other suitable situations. Their foraging areas are near woodland and watercourses where insect food is abundant. Most bats remain within an 80 km radius of the cave with a maximum recorded distance of 275 km. In winter they return to the cave, which is providing a suitable adverse season habitat, but in summer it is improbable that sufficient food would be available for so large a population in its immediate vicinity. Longer return movements have been recorded or inferred for the hoary bat (*Lasiurus cinereus*), which occurs in Alaska and northern Canada in summer and in winter is mainly found south of latitude 37°N, and in *Myotis sodalis*, which hibernates in caves in Kentucky and spends the summer up to 560 km north in Indiana, Ohio and Michigan. Ecological explanations of many long-distance bat migrations are still awaited, and the problem is complicated by the apparent inconsistent migratory behaviour of some species. This is exemplified by the guano bat (*Tadarida brasiliensis*). After weaning the young, the bat flies from Arizona and Texas to Mexico, travelling up to 800 km within two weeks. The return journey to the nursing colonies is made in spring. However, not all the bats behave in this way. A substantial proportion do not move south to Mexico, but hibernate in northern caves, while others do not return from Mexico in spring but remain to breed there (Yalden and Morris, 1975).

One of the most spectacular examples of migration is witnessed in the ungulates of East Africa, where the wildebeest alone number more than 800 000 (Delany and Happold, 1979; Pennycuick, 1975). They are accompanied in this migration by zebra and the small Thomson's gazelle. Each species has its characteristic dietary requirements. Zebra consume considerable quantities of the tall and nutritionally poor parts of grasses. Wildebeest eat short grasses, taking a large proportion of leaf and a small amount of stem. These are of a nutritionally higher quality than the grass consumed by the zebra. As a result wildebeest consume, for their size, relatively less bulk of food. Thomson's gazelle eat short grasses and herbs and, of the three, have the greatest demand for high quality food (section 4.5). These trophic preferences result in a sequential use of the grasslands with zebra moving in first, followed by the wildebeest and finally the gazelle.

The Serengeti plains (Fig. 6.7) have a low rainfall (514 mm per annum) and produce a rapid growth of short high-protein grasses during the wet season from November to May. They then attract many animals. At the end of the rains the productivity of the plains falls and the animals have to go elsewhere. They move to the western Serengeti, where a higher rainfall (976 mm p.a.) and a longer wet season provide food after the supplies of the plain have been exhausted. There is also considerable riverine vegetation within this area. Finally, in August and September, the herds take up residence in northern Serengeti where the rainfall is still higher 1100 mm p.a.) and the grasses have a longer growing season, assisted by a higher soil moisture content. In this clockwise migration the species interactions are of interest. Trampling and removal by zebra makes the grasses more suitable for consumption by the wildebeest. Their intense grazing stimulates the new plant growth preferred by the gazelles and makes the smaller herbs more accessible. This migratory grazing system is

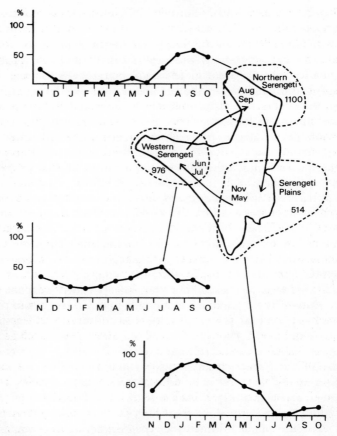

Figure 6.7 Migration of wildebeest in Serengeti. Annual rainfall figures (mm) and monthly percentage of wildebeest present are given for each sector of the Park (modified from Delany and Happold, 1979).

highly adapted to the different productivity levels and rainfall regimes of grasslands in juxtaposition. The ungulates make maximum and subtle use of each of the component areas, the higher rainfall locations in the west and north permitting particularly intensive use of the low rainfall plains.

6.5 Quantitative description of niche

Much of the foregoing account emphasizes differences between species' ecological requirements for a single variable, e.g. food, space, time. This was in several instances found to provide an inadequate definition of niche, and, in accordance with the concept of the fundamental niche, reference was made to additional variables. Even so, there is scope for more elaborate and quantitative analysis in which account can be taken of the extent of use of each resource. The niche breadth of a species can be expressed for a single resource with a higher value indicating high use of the different components of that particular resource. This type of analysis has been undertaken on the five commonest species of rodents inhabiting the Nossob River valley of the southern Kalahari Desert. Four species occupied the low dunes, riverbed plateau, and high dunes comprising the area, but not necessarily with the same frequency; the fifth (*Desmodillus*) occupied all these habitats except the riverbed. Populations were monitored from 1970 to 1976 and food and habitat niche breadths measured for summer and winter (Table 6.2). Seasonal changes occurred in niche size. In 1972, the habitat niches were narrower in summer when populations were lower. The results on feeding are less uniform. *Rhabdomys* and *Mus* have greater ranges of foods in winter than in summer; for *Gerbillus* and *Desmodillus* the reverse holds, and for *Tatera* there is little change throughout the year. Measurements of niche overlap between pairs of species pointed to a greater overlap in both habitat and food in winter. These analyses are important in focusing attention on niche dimensions and overlaps for two factors, although others, such as activity and size, should not be undervalued.

An alternative approach has involved measurement of the extent of utilization of different habitat components by each species within a common area, and then subjecting the data to a multivariate analysis such as discriminant function analysis. This has been undertaken for four species of small rodent inhabiting oak–hickory, chestnut–oak and pine forest woodland in eastern Tennessee. Eight habitat variables were

Table 6.2 Niches of Kalahari rodents (after Nel, 1978)

	Size (g)	Activity	Relative density	Niche breadth			
				Habitat (72)		Food	
				W	S	W	S
Gerbillurus	25.9	N	3.31	0.60	0.48	0.52	0.89
Desmodillus	46.1	N	0.37	0.36	0.10	0.57	0.87
Tatera	64.9	N	0.37	0.10	0.03	0.63	0.69
Rhabdomys	32.0	D	4.68	0.47	0.16	0.75	0.59
Mus	4.7	C/D	0.37	0.19	0.05	0.49	0.33

N–nocturnal, D–diurnal, C–crepuscular, W–winter, S–summer

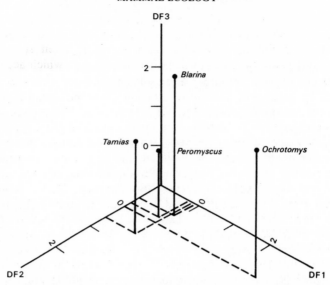

Figure 6.8 Multivariate analysis of niche occupancy by forest-floor small mammals in eastern Tennessee for three most important discriminant functions (DF) (modified from Dueser and Shugart, 1979).

identified, including vertical density of herbage, shrub-level density of vegetation and the evergreenness of overstorey. The four rodents occurred in each habitat type, and the conditions and frequency of trapping at each trap position was recorded. The results of this analysis (Fig. 6.8) indicate a separation of the species from each other, with *Ochrotomys* particularly distinct. This technique analyses all the variables simultaneously, and attempts, through elaborate analysis of correlations between habitat variables, to reduce the description of the habitat to a smaller number of parameters, or discriminant functions (DF). In this analysis, a discriminant function incorporates several habitat variables, although all are not of equal importance: the predominant variable in DF 1 is evergreenness of the overstorey, and less important are vertical foliage in woody vegetation and density and size of tree stumps. The trend within DF 1 is from evergreen to deciduous overstorey. DF 2, which accounts for 20 % of the variation, describes shrubbiness of vegetation, with *Tamias* at the sparse end of this measure. Although useful in simplifying habitat description, multivariate methods do no more than analyse existing data and do not necessarily identify the features of the habitat which are most important to the species in question.

CHAPTER SEVEN

ENERGETICS

Mammals are primary, secondary or tertiary consumers within ecosystems. Their effects, in energy terms, on an ecosystem depend on individual energetic requirements and the number of animals present. For present purposes it is convenient first to consider how energy is utilized by the individual (i.e. its energy budget), and then how populations of mammals influence and make use of the energy resources available to them. While this chapter restricts itself to energy relationships, it is also possible to consider the flow of chemical elements, many of them essential to animals, through the system in the same way.

A. Energy Budgets

The course of food energy is outlined in Fig. 7.1, with figures included in

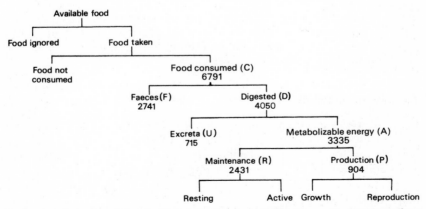

Figure 7.1   The scheme of energy flow through an animal. The figures refer to a sheep (from Blaxter and Graham, 1955).

99

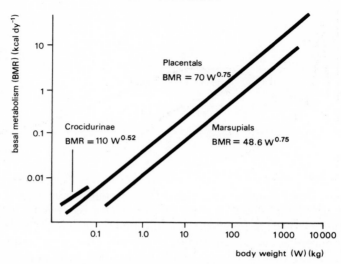

Figure 7.2 Relationships between weight and metabolic rate in mammals (modified from Dawson and Hulbert, 1970; Kleiber, 1961; Vogel, 1976).

kilocalories for the daily energy budget of a sheep. (Energy can also be measured in joules: 1 calorie = 4.18 joules). All the figures balance:

Consumption (C) = faeces (F) + digested energy (D)
D = excreta (U) + metabolizable energy (A)
A = energy used in respiration (R) + production (P).

When determining energy consumption it is often essential to obtain some data from laboratory studies. Furthermore, some parameters are more readily measured than others, and as a result, those less easy to obtain are derived from these formulae by subtraction.

7.1 Basal metabolism

Basal metabolism is the resting metabolism of an animal when there is no food in the stomach, and at a temperature which involves no energy expenditure for heating or cooling. It can be measured through O_2 consumption and CO_2 output or through heat production. In placentals the relation between O_2 consumption (as a measure of metabolic rate) and body weight is expressed by

$$V_{O_2} = 4.0\,W^{-0.25}\,\mathrm{cm^3 \cdot O_2\,g^{-1}h^{-1}}$$

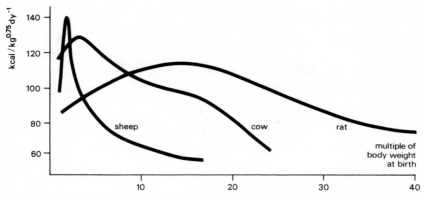

Figure 7.3 Relationship between basal metabolic rate (expressed as kcal/kg$^{0.75}$ dy^{-1}) and relative growth as multiple of mean initial body weight (modified from Poczopko, 1979).

where W is weight in grams; for marsupials the relationship is

$$V_{O_2} = 2.6\,W^{-0.25}\,cm^3 \cdot O_2\,g^{-1}h^{-1}.$$

These are alternatively expressed in kilocalories per day, with weight in kilograms. The *Basal Metabolic Rate* (BMR) for placentals is BMR (kcal dy^{-1}) = $70\,W^{0.75}$, and for marsupials BMR (kcal dy^{-1}) = $48.6\,W^{0.75}$ (Fig. 7.2). Thus as mammals increase in weight their energy requirements become relatively less. Most mammals conform to these formulae, but there are exceptions. Placental fossorial mammals of less than 60 g have a higher metabolic rate than these formulae would suggest, and those of more than 80 g, a lower one (McNab, 1979). The shrews have appreciably higher metabolic rates. For the Soricinae BMR is $167\,W^{0.67}$ and for the largely tropical Crocidurinae it is $110\,W^{0.52}$ (Fig. 7.2); the Neomyinae and Blarinae occupy intermediate positions.

Basal metabolic rate, and hence energy utilization, can vary in relation to numerous factors even after account is taken of increases in weight. In some species there is a relative increase in BMR during early development with the pattern far from uniform (Fig. 7.3). In the sheep the high BMR lasts for approximately only the first nine weeks of life and then rapidly declines, whereas in the rat the build-up to a peak takes seven weeks and a further six weeks to gradually approach the more normal level. The number of young in a litter can determine the amount of energy used, probably through the effects of huddling. In the bank vole (*Clethrionomys*), individual energy demand is highest in litters of one or two and decreases as litters increase in size (Gebczynski, 1975).

7.2 Metabolism under natural conditions

Basal metabolism applies to an animal under a narrowly defined set of conditions. In the wild the animal has periods of rest, feeding and activity as well as experiencing a range of temperatures. The maintenance energy demand varies from species to species according to size and pattern and type of activity. For example, a species which is huddled in a nest at night and active during the day is likely to have smaller energy demands than a species which is active at night when temperatures are lower. There is also the energy required for reproduction and growth. The total can be expressed as

Total energy required dy^{-1} = basal metabolic energy + energy of activity

+ production energy + energy of heat maintenance.

The activity energy can be further broken down to consider the different energy demands of, for example, resting, standing, feeding, walking, running and breeding.

A simple example of daily energy expenditure of maintenance is provided by the pocket mouse *Perognathus parvus* (Fig. 7.4). This takes account of the time spent active, resting, and torpid, as well as the prevailing temperatures at different times of the year. From this type of

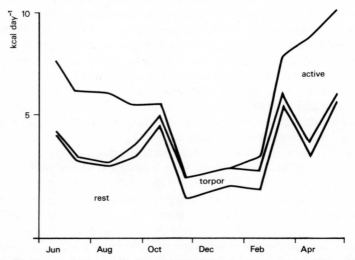

Figure 7.4 Estimated daily maintenance energy expenditure of male Great Basin pocket mouse (*Perognathus parvus*) (from Schreiber, 1978).

information on this and other species it is possible to obtain the *Average Daily Metabolic Rate* (ADMR) which is a measure of the mean daily energy consumption. For small mammals, this has usually been obtained from laboratory simulations of field conditions where it has been found that a correlation exists between ADMR measured in $cm^3 O_2 g^{-1} hr^{-1}$, and body weight (g). For grassland rodents

$$ADMR = 19.94 W^{-0.50}$$

or

$$ADMR \, kcal \, g^{-1} \, dy^{-1} = 2.297 W^{-0.50};$$

for insectivores

$$ADMR \, cm^3 \, O_2 \, g^{-1} \, hr^{-1} = 26.80 W^{-0.50}$$

or

$$ADMR \, kcal \, g^{-1} \, dy^{-1} = 3.087 W^{-0.50}$$

(French *et al.*, 1976). An extension of ADMR is the *Daily Energy Budget* (DEB) which also takes account of energy used in thermoregulation and reproduction.

For large mammals, energy costs have been measured for different field activities, and from the time spent in each activity the total metabolism for the day has been calculated. Basal metabolism and activity cost can be expressed as a multiple of BMR for each activity. Examples, for an animal weighing 100 kg, are: standing 1.1, running 8.0, walking on level 1.64, foraging 1.59, playing 3.0 and ruminating 1.26. These values vary with weight of animal, intensity of activity (e.g. running or trotting), and time of year. Examples of energy expenditure are given in Table 7.1. Further variations can result in increased energy of production demands of growth, gestation and lactation. For all mammals, *Yearly Energy Budgets* (YEB) are calculated from the approaches outlined above, and from these, annual population budgets can be obtained.

7.3 Consumption, digestion and assimilation

Relative to their size, small mammals have the highest consumption of food and large mammals the lowest. Intake is also influenced by digestibility and quality as well as the physiological state of the animal. The ratio of weight of daily food consumption to body weight in shrews typically ranges from 0.5 to 2.5, while for large mammals such as buffalo, hippopotamus and elephant the figure is about 0.01. Comparing the food

Table 7.1 Daily energy expenditure (kcal dy⁻¹) by large mammals (after Moen, 1973)

| Activity | ♀ Elk (Cervus) 160 kg | | ♂ Elk* (Cervus) 327 kg | | White-tailed deer (Odocoileus) 75 kg | | | | Pronghorn (Antilocarpa) 45 kg | |
| | | | | | Summer | | Winter | | | |
	% time	kcal dy⁻¹	% time	kcal dy⁻¹	% time	kcal dy⁻¹	% time	kcal dy⁻¹	% time	kcal dy⁻¹
Basal metabolism	100	3149	100	5383	100	1784	100	1784	100	1216
Bedding	34	1083	19	1029	16	280	25	442	23	280
Standing	41	1428	46	2694	16	308	25	487	23	308
Eating	46	947	17	615	69	667	50	490	54	315
Ruminating	26	241	28	446	16	68	25	107	23	60
Walking	24	1310	18	1681	69	1708	50	2829	53	668
Running	0	0	<1	41	0	0	0	0	1	97
Breeding	1	53	18	2836	0	0	0	0	0	0
		5062		9342		3031		2829		1728
Multiple of BMR		1.61		1.74		1.70		1.59		1.42

* Rutting harem bull.
Total activities exceed 100% as more than one activity is undertaken simultaneously.

intake of a small and a large carnivorous species, the common shrew (*Sorex araneus*), which weighs 9 g, annually consumes 3540 g of worms, molluscs, spiders, harvestmen, beetles and a few other invertebrates (Pernetta, 1976). This is more than 393 times its body weight. In contrast, the cheetah (*Acinonyx*), weighing 60 kg, consumes 1320 kg of the 3300 kg of impala, gazelle and hartebeest that it kills each year (Kruuk and Turner, 1967). Here the consumption/weight ratio is only 22.

The coefficients of digestion (D/C) and assimilation (A/C) are indicators of the energy use mammals make of the food they consume. Examples include grazing rodents and lagomorphs (67 % digestion, 65 % assimilation coefficients) omnivorous rodents (77 %, 75 %), granivorous rodents (90 %, 88 %), small shrews (?, 90 %), carnivores (88 %, ?) (Grodzinski and Wunder, 1975), and ruminants approximately 50–70 % and 40–60 % respectively. The last group has relatively low efficiencies and carnivores and granivores relatively high ones. Of the energy consumed by ruminants, there is appreciable loss through gaseous products of digestion (methane) and heat of digestive fermentation (Maynard *et al.*, 1979). In ruminants variation in diet quality can considerably influence digestibility. Food-stuffs such as grasses, which contain a large amount of cellulose, have a lower digestibility than young herbs. In roe deer (*Capreolus*) the summer diet of browse and herbs has a digestibility coefficient of 70 % whereas in winter it never exceeds 40 %. Lower digestibility can be compensated by higher intake; the problem here is the slower passage of less digestible food through the gut. For much of the winter the deer's energy intake is marginally sufficient and from time to time recourse has to be made to reserves built up in summer (Drozdz, 1979).

7.4 Production

Production is the energy utilized for pre- and post-natal body growth, repair of tissues, fat formation and the provision of reproductive products and milk. It is thus the total new tissue produced within a specified time period. Post-embryonic growth has been described for many species of mammals and generally follows a broadly common pattern (Fig. 7.5). This comprises an early phase of rapid growth, followed by a second phase of slower growth (usually as the animal develops from sub-adult to adult). Some mammals experience negative growth as old age is approached. There can also be negative growth in winter when several species, from shrews to deer, lose weight.

The energy requirements of early post-embryonic growth of the

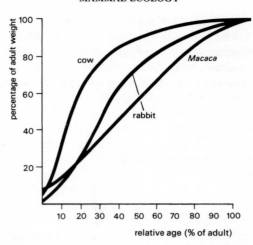

Figure 7.5 Standardized growth curves of cow (*Bos*), rabbit (*Oryctolagus*) and macaque (*Macaca*). From Altman and Dittmer, 1972.

laboratory mouse (*Mus*) (Fig. 7.6) show that soon after birth 25–30 % of assimilated energy goes into production, although by the time of weaning this has dropped to 10–15 %. Among domestic mammals, for a steer of 300 kg growing at the rate of 1 kg dy^{-1} the figure is 57 % (McDonald *et al.*, 1973). At the same time as energy demands are being made by the suckling young, considerable energy is required by the lactating mother. In *Mus* this is well in excess of the non-lactating female of the same weight (Fig. 7.6). In adult female deer the quantity of milk required depends on the age and activity of the fawn. Energy requirements may be as high as 2.30 BMR for females feeding twins at the peak of their demand (Moen, 1973). A dairy cow weighing 500 kg and producing 20 kg of milk expends 70 % of its assimilated energy on milk production.

Energy demands are less on the pregnant female than during the peak of lactation. Only for the last third of pregnancy is there a substantial increase in energy demand. In *Mus* (Fig. 7.6), as the weight of the embryo increases the total weight, the energy requirement per unit weight declines. For a deer weighing 60 kg, energy expenditure of activity is 1.42 BMR. This increases to 1.53 and 1.64 at the end of gestation with one and two fawns respectively.

Maintenance of tissues and laying down of fat frequently occur in adult animals. The quantity of assimilated energy used in this way is variable, particularly in domestic animals. Figures of 5 % have been cited for steers,

9.7% for red deer and 37% for sheep. Weight loss results in negative production. The souslik (*Citellus pygmaeus*) can lose as much as 62% of its weight during hibernation (Abuturov and Kuznetsov, 1976). White-tailed deer (*Odocoileus virginianus*) show losses of up to 23% from early November to early January; the largest declines are witnessed in males that have been in the rut (Moen, 1973).

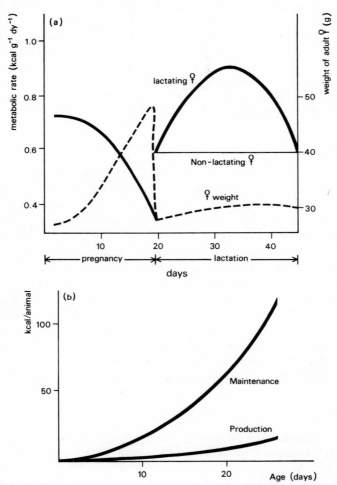

Figure 7.6 Metabolism of laboratory mice (*Mus*) in perinatal period. (*a*) Variation in metabolic rate of adult female; (*b*) daily energy requirement of young animal for assimilation and production (modified from Myrcha, 1975).

Small mammals grow faster than large mammals and convert a higher proportion of the food they consume to production energy. This is seen in the following comparison. A half-grown impala (*Aepyceros*) weighing 20 kg and supplied with its own weight of consumable food, will eat $0.4 \, kg \, dy^{-1}$ so that within 50 days it will have exhausted its supply. As its weight gain will be $0.1 \, kg \, dy^{-1}$ it will over this period have increased its body weight by 25%. A half-grown elephant (*Loxodonta*) of 1000 kg and provided with its own weight of food, eats $17 \, kg \, dy^{-1}$ and gains $0.3 \, kg \, dy^{-1}$. The food will take 58.8 days to be eaten and in that time there will be a weight increase of 17.6 kg or 1.76%. Thus for every kg of food consumed there is a daily addition of 5 g weight in impala and 0.3 g in elephant. These figures can vary with food quality, but even so the example highlights differences in conversion of food to flesh in different-sized mammals (Delany and Happold, 1979).

B. Energetics of populations

Having considered energy utilization in individual mammals we can now examine the routes energy takes in its flow from the vegetation into and through the natural populations made up of these individuals. This analysis involves the description in energetic terms of the available food, the quantities consumed (as well as that removed and not consumed) and the resulting production within the animal population. These are usually expressed to cover a period of one year and are for a specified area. These assessments thus represent an amalgamation of the energy demands of individual animals under natural conditions and the population dynamics of the species in question. Consideration can be given to the entire range of mammal species within an area, a selection of species, e.g. ungulates or rodents, or individual species. As information is frequently fragmentary and has often been accumulated for only a limited number of mammal species, opportunities to present a comprehensive picture are few.

7.5 Biomass

Biomass is a measure of the weight (mass) or standing crop of living tissue. It is usually expressed as wet weight for a defined area and is useful as an indicator of the quantity of animal or plant material supported. While it can be used as a measure of the total amount of organic matter present, it is more commonly used with reference to specified groups of plants and animals. By itself it does not permit many conclusions to be drawn about energy consumption as this is much influenced by the size of the animals (section 7.4) and their rates of turnover. Nevertheless, many estimates of biomass are available which provide useful insights into supportable levels of mammals.

Among small rodents, rapid population turnover and adaptations to changing conditions can result in substantial seasonal and annual changes in biomass. The level of support can also vary greatly between habitats. In North America the ungrazed tall grass (*Andropogon–Panicum*) supports an appreciably higher biomass than the mid-grass (*Agropyron–Stipa*) (Fig. 7.7). The moist tropical grassland (*Imperata–Cymbopogon*) maintained a consistently high biomass. These grasslands support up to 260 kg km^{-2}.

Among the cyclically fluctuating rodents changes are rapid and peaks high. In the Arctic tundra, estimates for the lemming (*Lemmus trimu-cronatus*) range from troughs of 0.4 kg km^{-2} to 400 kg km^{-2} (Batzli, 1975). It is possible that the latter figure may be appreciably exceeded by *Lemmus obensis* in northern USSR as well as by *Microtus* in Polish grasslands (Gromadzki and Trojan, 1971). Seven years of continuous observation in tropical rain forest in Nigeria revealed a biomass of ground-dwelling species of 23 to 123 kg km^{-2} (Happold, 1977). Unlike these

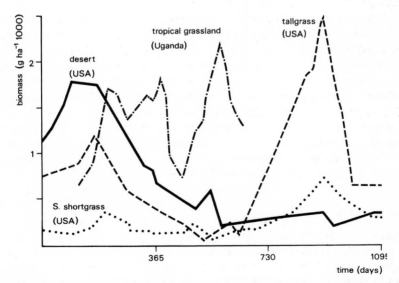

Figure 7.7 Standing crop biomass of rodents in three temperate and one tropical grassland site in North America and Uganda respectively. (The tall grass site also includes a species of shrew) (modified from French *et al.*, 1976, Cheeseman and Delany, 1979).

figures, estimates from Panama forests of 402 and 630 kg km^{-2} and from Malaya of 84 kg km^{-2} included large rodents and squirrels (Fleming, 1975). Undoubtedly one of the consistently highest estimates of small rodent biomass is from mixed bush and grass (*Pennisetum*) in eastern Zaïre. Here estimates made at various times of year ranged between 1115 kg km^{-2} in grass bush to 1646 kg km^{-2} in mixed elephant grass and bush (Dieterlen, 1967).

Biomass estimates of large herbivores have been undertaken extensively in the African savannas. Populations, and hence biomasses, are generally more stable than are those of rodents. Among the highest estimates are those from Kivu Park, Zaïre (17 448 kg km^{-2}); Rwenzori Park, Uganda (19 928 kg km^{-2}); and Manyara Park, Tanzania (19 189 kg km^{-2}). These range down to 984 kg km^{-2} in northern Kruger Park, South Africa, and 405 kg km^{-2} in the drier parts of eastern Kenya (Coe *et al.*, 1976).

Estimates of standing crop biomass have been obtained for all the living components of a small number of temperate grassland sites in North America. These provide useful comparative data on the position of mammals within these ecosystems. In the northern short grass (*Buteloua–Buchloe*) of Pawnee, Oklahoma, the total above- and below-ground biomass of primary producers was 2 280 000 kg km^{-2}. The biomass of soil microorganisms was 67 000 kg km^{-2} while the invertebrates and birds totalled 622.7 kg km^{-2} and 5.5 kg km^{-2} respectively. The mammals comprised 1119.3 kg km^{-2} of which cattle made up 1090 kg km^{-2}, small mammals 8.1 kg km^{-2}, pocket gophers 13 kg km^{-2} and antelopes 8.2 kg km^{-2}. Apart from the cattle, the mammal contribution is relatively small. In a mixed grassland at Matador, Saskatchewan, from which cattle were absent, the vertebrate above-ground biomass accounted for as little as 0.004% of the total (Coupland and Van Dyne, 1979). There was no indication that small mammals compensated for loss of cattle. Generalizations are difficult concerning the relationship between large and small mammal biomass at different sites. Undoubtedly the character of the vegetation is most important. In some natural moist tropical grasslands such as Rwenzori Park, Uganda, it is evident that high rodent biomass is associated with relatively low large mammal biomass (Delany, 1964). The converse also holds.

7.6 Production

Annual population production is obtained by measuring the growth increment of each individual in a twelve-month period. This must take account of new additions and the growth put on by animals that survive through, and die during, the year. Population production can be expressed as either mass (kg) or energy (cal) per unit area. To calculate this figure, the population structure and survivorship at different age intervals must be known. An example is provided by the African elephant (Table 7.2). As the total number in each age class is given within the specified time period, the table represents records from 10 933 elephant, or if account is taken. of deaths within the period, 10 487.5. These elephants had an average weight of 2287 kg. This results in a weight of 2287 kg × 10 487.5 = 23 984 912 kg for the whole life table population. During the year it increases its weight

Table 7.2 Elephant population production, Rwenzori Park, Uganda. (After Petrides and Swank, 1966)

(1) Age x	(2) No. alive at beginning of age x	(3) Median no. alive	(4) Weight average (kg)	(5) Weight increment (kg)	(6) Population weight increment (kg)(3) × (5)
1	1,000	850.0	91	91	77,350
2	700	630.0	204	113	71,190
3	560	532.0	318	113	60,116
4	504	478.5	454	136	65,076
5	453	430.5	612	159	68,450
6	408	387.5	794	181	70,138
7	367	348.5	998	204	71,094
8	330	313.5	1202	204	63,954
9	297	282.0	1406	204	57,528
10	267	260.5	1610	204	53,142
11	254	247.5	1814	204	50,490
12	241	235.0	2019	204	47,940
13	229	223.5	2200	181	40,454
14	218	212.5	2381	181	38,463
15	207	205.0	2586	204	41,820
16	203	201.0	2790	204	41,004
17	199	197.0	2990	204	40,188
18	195	193.0	3175	181	34,933
19	191	189.0	3379	204	38,556
20	187	185.0	3583	204	37,740
21	183	181.0	3742	159	28,779
22	179	177.0	3856	113	20,001
23	175	173.0	3946	91	15,743
24	171	169.5	4037	91	15,425
25	168	166.5	4082	45	7,493
26–67	3,107	3,026.5	4082	0	0
Total	10,933	10,487.5			1,157,067

by 1 157 067 kg. The production (P)/biomass (B) ratio is 0.048 for the population with this structure. The average production of each elephant was 110.3 kg yr^{-1}. In this locality the population density was 2.077 km^{-2} with annual production 229.1 kg km^{-2} yr^{-1}. The calorific value of elephant is approximately 1500 kcal kg^{-1} of live weight so that the energy production of tissue amounted to 343 650 kcal km^{-2} yr^{-1}.

In field studies of this type, a general calorific value of the live weight is used which takes account of differences between individuals within the population. For most mammal populations, this ranges between 1.5 and 1.8 kcal g^{-1}. Examples from North American rodents include *Perognathus* (1.55), *Onychomys* (1.61), *Peromyscus* (1.56) and *Clethrionomys* (1.69). Shrews have similar values, with 1.71 and 1.75 recorded for *Blarina* and *Sorex* respectively. Exceptions occur—*Zapus* has values in excess of 2.1. The energy content of an individual alters as the proportions of fat, water and muscle change. The energy in fat (8.7–9.2 kcal . g^{-1}) is approximately twice that of lean meat (4.4–4.8 kcal . g^{-1}). In laboratory and domesticated animals laying down of fat is common. Here, the proportions of fat can also alter as the animal develops. In newly born *Peromyscus*, fat comprises 7.4% of body weight, rising to 36% at 42 days. During this time caloric density increases from 0.89 kcal . g^{-1} to 2.23 kcal . g^{-1} (Kaufman and Kaufman, 1975).

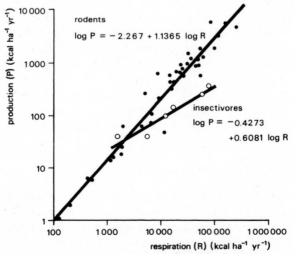

Figure 7.8 Population production as a function of respiration in rodents and insectivores (modified from Grodzinski and Wünder, 1975).

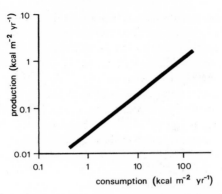

Figure 7.9 Annual population production as a function of consumption (modified from McNeil and Lawton, 1970).

The P/B ratio of the elephant population is close to 0.05, this figure probably having general application to large herbivorous mammals weighing more than 800 kg. For those weighing 100–750 kg and 5–90 kg the corresponding estimates are 0.20 and 0.35 respectively (Moen, 1973). For small rodents and insectivores population production can be obtained from respiration (Fig. 7.8). There is also a positive correlation for large and small mammals between consumption and production (Fig. 7.9). Estimates of production (Table 7.3) for selected groups of mammals show, as might be expected, considerable variation for similar groups, e.g. large herbivores, from one locality to another. From these examples it appears that small rodents do not attain production levels approaching those of large mammals in their most favoured habitats. These however are in the tropics. When comparisons are made in temperate and sub-arctic situations the discrepancies are not so great, and it is probable that in some arid areas the production by small mammals is higher.

7.7 Availability and consumption of food

The *Net Primary Production* (NPP) is the total plant material produced and is potentially all available for consumption by animals. The *Available Primary Production* (APP) is the plant food accessible to the species in question and falling within its dietary choice. Thus what is taken depends on dietary habits and mobility. Species such as cattle living in grasslands are often able and prepared to consume most of the above-ground

Table 7.3 Annual population production estimates from various localities

Species (1)	Location (2)	Annual production	
		$kg\,km^{-2}\,yr^{-1}$ (3)	$kcal\,km^{-2}\,yr^{-1}$ (4)
Bison, moose, elk	boreal aspen forest-grassland, Canada	1100	
Large herbivores	Manyara Park, Tanzania	2405	
Large herbivores	arid, Kenya	87	
Large herbivores	Serengeti Park, Tanzania	1743	
Large herbivores	savanna, Zimbabwe	684	
Beef cattle	N. American grassland (hypothetical)	600	
White-tailed deer	George Reserve, Michigan	400	
Saiga	USSR steppes	100	
Three rodents	*Picea* forest, Sweden	315	472 036
All rodents	*Tilia* woods, Poland	136	198 295
All rodents	savanna grassland, Uganda	550–700	
Small mammals	ungrazed tallgrass, USA		322 500
Small mammals	desert grassland, USA		182 200
Small mammals	midgrass, USA		19 900

Approximate conversions of these figures can be made from column (3) = column (4)/1500.

vegetation in their habitat and so have an APP/NPP ratio approaching unity or 100%. In contrast, forest-dwelling small rodents and ungulates may have low proportions of the total production available as much of the production will be out of reach in the canopy; figures of 4 to 7% have been suggested as typical for the rodents (Hayward and Phillipson, 1979). Granivores must also have low values as seed production represents a small proportion of the net primary production.

There are further considerations in connection with food availability. Annual fluctuations in production can be considerable. Years of good fruiting and seed production in temperate woodlands can result in the provision of excellent winter food resources. A good year of beech mast production can result in a 23-fold increase of the ground litter input over a bad year. Years of low and high rainfall in tropical and sub-tropical situations can also considerably influence levels of primary production. Within all seasonal habitats there is the added complication of high production, and hence availability at some times of year and low or zero production at others. The frequent result is periods of excess followed by periods of shortage. This is well illustrated by the large and small herbivores inhabiting the tall grasslands (*Themeda, Pennisetum,*

Sporobolus) of the Serengeti Plains of Tanzania. Here there are 12 species of large mammals, the commonest being wildebeest, zebra and Thomson's gazelle. The small ones are mainly represented by *Otomys* and *Praomys*. During the wet season, from November to June, new food is being added each month (Table 7.4). Some is eaten, but what is not is still nutritious and palatable and can be added to the following month's available supply. This is not the case in the dry months when no new food is added and the June supply of grass dries out by the end of the month. The fluctuation in ungulate demand is due to migration of some animals into and out of the area. Thus while the total annual production of vegetation of 5978 kg ha^{-1} yr^{-1} is appreciably in excess of the mammal consumption of 1191 kg ha^{-1} yr^{-1} (19.9%), the seasonality of its appearance means that the supply is inadequate at certain times. Considerable quantities of grass are lost in burns (53%), or consumed by detritus feeders (19%) and 38 species of grasshoppers (8%).

Similar estimates of the proportion of primary production consumed were obtained for *Acacia* savanna in Tsavo Park, Kenya, from 1969 to 1972. Here, elephant featured prominently within the fauna, and during

Table 7.4 Mean monthly food availability and requirement (kg/ha) in the tall grasslands of Serengeti National Park. (After Sinclair, 1975.) Invertebrate food requirement is not included.

Month		Monthly production	*Food requirement*			Available food	Remaining food
			Ungulate	Small mammal	Total		
Nov.		555	111.33	3.33	149.56	598.8	440.3
Dec.		781	29.22	8.87	94.59	1277.8	1183.3
Jan.		668	45.04	8.87	110.41	1907.8	1797.3
Feb.		633	48.07	8.87	113.44	2486.8	2373.4
March		857	48.07	8.87	113.44	3286.9	3173.4
April		1154	48.07	4.95	109.52	4384.0	4274.4
May		706	43.95	4.95	105.40	5036.9	4931.5
June		340	129.50	4.95	190.95	5238.0	5137.6
July	⎫	0	124.60	4.95	139.55	0	0
Aug.	⎬ Dry	37.5	164.60	3.33	173.83	38.4	0
Sept.	⎨ season	92	164.60	3.33	172.53	92.9	0
Oct.	⎭	154	164.60	3.33	172.53	154.9	0
Totals		5978	1121.65	68.60	1645.25		

Note: During wet months (Dec.–June) grass that remained uneaten at the end of the month has been added to the production of the subsequent month. During the dry months (July–Oct.) any remaining grass dried rapidly and was unavailable in a subsequent month.

this period there were considerable losses, possibly 25 % of the population, as an indirect result of drought. The estimated consumption by mammals ranged between 9.5 and 18.2 % of the above-ground primary production (Phillipson, 1975). Temperate ungulates encounter similar problems of seasonality of supply. The summer food in European woodlands is sufficient to support a population of roe deer (*Capreolus*) seven times that which can be supported in winter (Drozdz, 1979).

The situation in pasture ecosystems can be very different. In meadows in France, cattle consume 85 % of the energy contained in the net above-ground primary production, while in a sheep-grazed pasture in Poland the figure is 86 % (Haas and Ricou, 1979). It has been demonstrated experimentally in Australia that a three-fold increase in stocking level of sheep from 10 to 30 ha^{-1} results in the sheep increasing their take from 37 to 58 % of annual shoot production (Breymeyer and Van Dyne, 1978).

The annual consumption of available primary production by small mammals has been obtained for several temperate habitats. These figures are generally low, and range from 0.50 to 5.20 % in grasslands, 5.50 to 8.73 % in deserts and 0.60 to 4.60 % in forests (Hayward and Phillipson, 1979). As these figures represent the amount of available food taken, the gross offtake must in many cases be extremely small. Exceptions occur at times of cyclical peaks. In arctic ecosystems, lemmings have consumed up to 70 % of the moss in Finnish Lapland and all the monocotyledonous material in Alaska (Batzli, 1975). Further south, in more temperate grasslands, up to 35 % of available food has been taken by microtines.

In most of these natural systems the energetic requirement of mammals is apparently relatively modest when compared to invertebrates. There are, though, general features that merit attention and place the mammal role in better perspective. At certain times, particularly when their demands are approaching or exceeding available resources, mammals can have considerable impact. The maintenance energy costs of mammals are relatively high compared to invertebrates. Among the grassland invertebrate herbivores, 9 to 25 % of ingested energy is converted to animal tissue, whereas from the same ecosystems the figures for sheep and cattle are 3 to 15 %. Thus the long-lived mammals put greater energy into maintenance and less into production and turnover. It is evident that habitats and their mammal populations can be manipulated to maximize production of selected species, as with cattle and sheep, although in natural habitats there are typically larger numbers of species having a proportionately lower energy demand.

There is little information on consumption relative to availability by insectivores and small carnivores. In an old field in Michigan, the least weasel (*Mustela rixosa*) annually consumed 31 % of the annual production of meadow mice (*Microtus pennsylvanicus*) (Golley, 1960), while in Wytham Wood, Oxford, weasels (*Mustela nivalis*) annually consumed 14.2 % of the small mammals (Hayward and Phillipson, 1979). In the fluctuating cycles of small rodents, the predatory impact varies within wide limits. At times of low densities, the predators may be taking a relatively high proportion of the herbivores and at times of high densities they may have only a limited impact. This is found in meadow mice (*Microtus californicus*) in California where up to 88 % were removed by predators at times of low density and only 5 % at peak densities (Pearson, 1964).

For large carnivores, estimates are available from the African savanna of their densities and biomass as well as those of their food sources. Here a guild of five predators comprising lion, leopard, cheetah, spotted hyaena and hunting dog feed on several ungulates including wildebeest, gazelles, zebra and buffalo (Schaller, 1972). Estimates of total predator biomass in five localities (Ngorongoro, Manyara, Serengeti, Nairobi, Kruger) range from 1:301 to 1:94 of the prey biomass. Although the energetics of these relationships have not been obtained, it is known that several factors combine to account for these differences. The highest ratio (1:301) is from Serengeti, where herbivore migrations greatly influence their availability, and the relatively high ratio (1:174) from Manyara is in a locality of high elephant populations. Here production by these largest of mammals is relatively low (section 7.4) and they are less easy to catch than smaller herbivores. The remaining three localities fall within the predator–prey biomass ratios of 1:94 to 1:108.

More quantitative data are available from Isle Royale (section 5.6), where there is a relatively simple predator–prey relationship between moose and wolves in which both populations have attained relative stability (Jordan *et al.*, 1971). Here the mean monthly moose biomass is 6.76 kg ha^{-1} and the production 1.67 kg ha^{-1} yr^{-1}. The animals typically available to the wolves, i.e. calves and adults more than six years old, amount to 1.24 kg ha^{-1} or 18 % of the standing crop. Of this the wolves consume 0.76 kg ha^{-1}. The wolf biomass is 1.7 % that of the moose; it consumes 11.2 % of the standing crop and 45.5 % of the production. This informative example illustrates how a small number of predators is supported by a relatively large number of prey, and furthermore, even if the hunting strategy of the wolves were more efficient, there still remains potentially more available food for utilization without necessarily reducing the moose population.

7.8 Habitat consequences of consumption and removal

The major impact mammals have on their habitat is through the modification of vegetation. This is frequently as a result of their normal feeding activities in the course of which the vegetation is steadily removed. Typically, a balanced relationship exists between the herbivore and its food source, and if the former is removed the vegetation will react to the alleviation of this pressure. Considerable interest has focused on mammal–vegetation interaction, particularly in the context of manipulation of their populations. This has relevance to maximization of yield of large mammals and the control of pest species. The following examples illustrate the closeness of plant–herbivore association and the ways in which mammals contribute directly to the detritus component of the energy cycle by the addition of plant material they do not consume. Mammals do modify their habitats in other ways. Examples include construction of burrow systems and consequent movement of soil and more ready penetration of the substratum by water, building of nests, use of tracks, and disturbance caused through courtship and mating.

In Britain, the rabbit (*Oryctolagus*) has modified heathland vegetation (Thompson and Worden, 1956). At the initial stage the rabbit consumes the ling (*Calluna*) around its burrow. This is replaced by the less favoured sand-sedge (*Carex*) and the rabbit moves further into the ling. As a result, the sedge area gradually increases. Eventually the rabbit has to travel too great a distance to the ling and the sedge is eaten to be replaced by bent (*Agrostis*) and fescue (*Festuca*) grasses. Heathland is thus replaced by grassland. On chalk grasslands, rabbits graze the plants close to the ground and reduce the variety of species that are able to survive. The most resistant plants to nibbling are those able to put out shoots on or close to the ground surface. As in heaths, the regeneration of shrubs and trees is prevented by rabbits, and the closeness of the grazing makes the vegetation unavailable to larger herbivores.

One of the most important mammals of the Central Asiatic deserts and semi-deserts is the small rodent *Rhombomys opimus* (Naumov, 1975). This species has extensive, elaborate and localized burrow systems. Around the burrows, intensive feeding results in removal of the vegetation. Slightly further away, in a broad belt of less intensive grazing, there is an influx of annual grasses and Cruciferae providing improved pasture for large mammals.

In cultivated plants there can be serious reductions as a result of damage and consumption. In Poland, *Microtus* daily consumed only 0.18 % of the yield of alfalfa, but effected a loss of 2.5 % through indirect effects on crop quality, e.g. poor growth and weed invasion (Grodzinski *et al.*, 1977). In East Africa periodic outbreaks of rodents (*Praomys, Rhabdomys, Arvicanthis*) have resulted in maize, wheat and barley losses through damage and consumption of up to 34% (Taylor, 1968). In young conifer plantations in Scotland, the consumption of bark by voles (*Microtus*) can result in complete destruction of the trees. In Russia, *Microtus* damage to maple (*Acer*) and elm (*Ulmus*) is confined to trees less than 123 cm tall, with most damage when they are 105 cm. Estimates of rodent damage to tree seedlings have been put at 80–100% destruction of oak (*Quercus*), elm (*Ulmus*), maple (*Acer*) and lime (*Tilia*), 50–60% destruction of ash (*Fraxinus*) and rowan (*Sorbus*), and 0–20% destruction of hazel (*Corylus*) and bird cherries (*Prunus*) (Sviridenko, 1940). In contrast, damage by small rodents to mature deciduous and coniferous trees is slight.

High seed consumption by rodents is common even when they are at low densities. Figures from habitats in America and Russia indicate consumption of conifer seeds at 22–60% of the crop. Grassland rodents also readily consume seeds. They can also be selective. In California, rodents consumed 75% of the seeds produced by *Avena*, even though this grass comprised only 4% of the plant cover (Hayward and Phillipson, 1979).

Extensive areas of upland Scotland support red deer (*Cervus*) and sheep (Mitchell *et al.*, 1977). Deer eat a higher proportion of ling (*Calluna*) and less grass than sheep. Areas stocked exclusively with deer revert to a dwarf-shrub heath. On the other hand, sheep stocking results in increased pressure on the ling which favours establishment of *Festuca* and *Agrostis* grassland. Many of these uplands would, in the absence of deer and sheep, revert to juniper (*Juniperus*), birch (*Betula*), rowan (*Sorbus*), and Scots pine (*Pinus*) woodland. Bark stripping of older trees by deer results in death and reduced tree production.

Experiments which for two years excluded elephant, hippopotamus and buffalo from small patches of woodland in western Uganda demonstrated striking changes in the vegetation (Spence and Angus, 1971). Grasses (*Sporobolus* and *Hyparrhaenia*) were reduced and the herbs increased, particularly one of the legumes (*Glycine*). Shrubs, notably *Acacia*, increased in numbers and stature, with a five-fold increase in those more than 60 cm tall. There were exceptions—one shrub (*Lonchocarpus*) showed a marked decline in numbers. As with the deer and sheep, man's interference with large mammal populations can greatly influence the habitats in which they live. It is further possible that such modifications can involve feedback mechanisms on the part of the vegetation that subsequently lower the carrying capacity of the habitat.

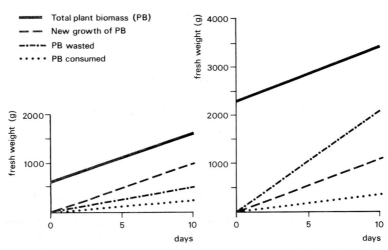

Figure 7.10 Green plant biomass consumed and wasted by a female *Microtus agrestis* in relation to new growth. Two consecutive 10-day feeding trials in outdoor enclosures (modified from Myllymäki, 1975).

Not all the vegetation removed by mammals is consumed. Frequently some parts fall to the ground to be incorporated into the soil. When *Microtus* was fed on naturally growing vegetation comprising two species of grass and red clover, there was a large quantity of wasted, unconsumed plant material (Fig. 7.10) which increased as the experiment proceeded. As plant material was removed, there was compensatory growth of the vegetation. The quantity wasted exceeded that consumed. Wastage also occurs when food is stored in caches and not subsequently recovered. While there is frequently considerable uncertainty to the quantity of such food recovered, there is no doubt that some animals set up considerable winter stores. The hamster (*Cricetus*) of eastern Europe, weighing 350 g, accumulates 15 kg of food such as maize seeds in its burrows between August and October (Gorecki and Grygielska, 1975). Squirrels are also notorious hoarders, storing nuts, mast and fungi in tree hollows, dreys or holes in the ground.

Destruction by large mammals can be exemplified by the elephant. In *Terminalia* woodlands, consumption of bark results in tree death and reversion to grassland. Elephant will also kill *Euphorbia* trees on which they rub themselves. In Rwenzori Park, 2.8 adult trees $yr^{-1} km^{-2}$ have been lost in this way. This loss would have to be made good by the $11\,058\,km^{-2}$ young trees in the area (Eltringham, 1980).

CHAPTER EIGHT

SOME APPLIED PROBLEMS

There are many ways in which mammals and man impinge, all too often with a resulting conflict of interests. As a result, man has often assumed a management role over mammals, which ranges according to circumstances from conservation through rational exploitation to elimination. The achievement of satisfactory ends frequently necessitates an adequate knowledge of the ecology of the species concerned. The following examples highlight particular problems, but should also be looked upon as having a much broader conceptual application. For example, the management and conservation of large terrestrial mammals is a worldwide problem, not uniquely African. Similarly, the conservation of many carnivores requires attention; the status of the otter in England (described here) illustrates the sort of problems confronting these animals. There is a further dimension. While the scientist can offer advice on the management of mammal populations, the ultimate decisions on their implementation frequently rest with administrators and politicians whose judgements are often much influenced by economic and social factors.

8.1 Game cropping

There are several ways in which wild terrestrial mammals can be used as an economic resource. They can, in their natural and semi-natural habitats such as National Parks, be the subject of tourist attraction. Elsewhere they can be the quarry of the sportsman. They can be semi-domesticated—this has been successful with eland (*Taurotragus oryx*) in the USSR (Treus and Kravchenko, 1968), while African species such as oryx (*Oryx*) (King and Heath, 1975) demonstrably have potential for use in this way. Finally, there is the possibility of cropping wild game animals to provide a

121

continuing resource. The arguments advanced in favour of this approach to wildlife management have placed considerable emphasis on the better adaptation of wild mammals, particularly with regard to dietary requirements and disease resistance, to many natural habitats than introduced domestic stock. As a consequence, wild mammal yields could be higher and the habitat not subjected to the same extent of modification and deterioration.

Theory

A basic premise of cropping a resource is the provision of a regular yield for an indefinite period—there must be what is called a sustained yield. Furthermore, the most desirable exploitation is for this to be maximal or a *maximum sustained yield* (MSY). To understand how this can be achieved, reference can be made to the logistic growth curve (Fig. 8.1). Here, the population growth curve stabilizes at the upper asymptote, this being the

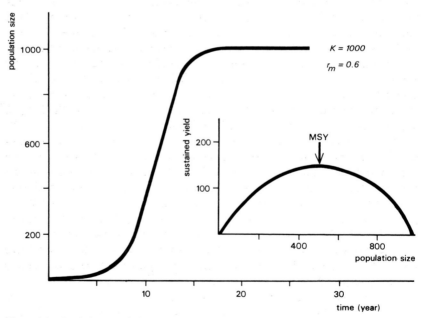

Figure 8.1 Logistic population growth and relationship between sustained yield and population size. MSY = Maximum Sustained Yield. K = population size at steady density (upper asymptote); r_m = intrinsic rate of increase; r = exponential rate of increase, N = population size and $r = r_m (1 - N/K)$ (modified from Ricker, 1958).

stage at which most natural populations exist. If the population is to be cropped, the phase of the curve at which this can be done most advantageously is at its point of inflection, when it is growing most rapidly. Expressed mathematically, the rate of harvesting must hold the population with an exponential rate of increase (r) at zero (Caughley, 1977). In this theoretical example (Fig. 8.1), to obtain a maximum sustained yield (MSY) the population must be reduced to $K/2$ which becomes the harvested population $\bar{N}$. The instantaneous harvesting rate (H) here is $r_m/2$ where r_m is the innate capacity for increase. The yield is then $\bar{N}H$. This results in the population being reduced to 500 ($\bar{N}$) and the MSY 150. In this theoretical example it is assumed that the population is being steadily removed and that natural losses and additions occur throughout the year. In practice, removal may take place over a limited part of the year, and may involve the maintenance of a sustained, as distinct from a maximum sustained yield, from a potentially increasing population. The instantaneous rate of harvesting (H) is the same as r, the exponential rate of increase. This is converted to the isolated rate of harvesting $h = 1 - e^{-H}$. It is this proportion of animals that can be taken at the time of harvesting. If the harvesting is spread over two or more periods of the year then $h = 1 - e^{-H/y}$, where y is the number of periods. This analysis highlights several practical issues. Firstly, cropping a population at its natural, relatively stable density can only produce a negligible sustained yield if the population is to be kept at that level. If the population is permanently reduced by approximately half there are likely to be effects on the habitat itself as a result of the modified feeding pressure.

For a population reduced from the steady state, there are two levels of density capable of producing the same sustained yield, but only one density that can produce an MSY (Fig. 8.1). A constant annual removal rate will result in the population stabilizing at the upper size for which that number is the sustained yield. If the annual removal exceeds the MSY, the population will ultimately become extinct. For many wild populations it is uncertain whether they are at their steady state, while intrinsic rates of increase are frequently not known. Even so, it is possible to derive these parameters by selective removal. Harvesting can itself be a device for revealing population characteristics.

Practice

Harvesting of large mammals has been practised in many parts of Africa and sustained yields have been obtained on numerous occasions. One of

Table 8.1 Sustained yield cropping of large mammals in Africa (after Field, 1974, 1979)

Location	Date	Species	Numbers	Meat dressed (kg)
Semliki, Uganda	1963–70	Uganda kob	6964	320,344
Acholi, Uganda	1965–67	Uganda kob	475	25,884
		Buffalo	368	92,365
		Hartebeest	116	9,735
Kruger Park, South Africa	1972–74	Elephant	1978	
		Buffalo	5432	
Loliondo, Tanzania	1969–71	Zebra	775	

the earliest experiments was on the Henderson Range in Rhodesia (now Zimbabwe) over 20 years ago (Dasmann and Mossman, 1961; Dasmann, 1964). Here it was proposed that 13 species of ungulate should be cropped. Of these three species, impala, zebra and warthog accounted for 80% of the total animals removed and 63% of the meat yield (kg). The percentages of their populations recommended for cropping were impala 25, zebra 20 and warthog 50. The same area was capable of producing, in addition to the game, an approximately equivalent amount of beef.

At the Akira range in Kenya, annual cropping rates of Grant's gazelle (*Gazella*), hartebeest, Thomson's gazelle (*Gazella*) and giraffe (Field and Blankenship, 1973b) have been recommended at 50, 35, 76 and 24% respectively, assuming optimal conditions of good nutrition, high survival of young, control of predators and low incidence of disease. The wide range of cropping rates take account of differences in age to maturity and reproductive rate. Other examples of sustained yield cropping appear in Table 8.1. In addition there have been numerous occasions when reduction cropping has taken place (Field, 1979). In East Africa these have involved hippopotamus, elephant, buffalo, Thompson's gazelle and impala. As a result the populations did not experience serious decline, and substantial quantities of meat were put on the market. In all these examples there is little evidence of practice matching theory, with the result that while sustained yields could be obtained, much less certainty surrounds the achievement of MSY.

The need for a sustained yield or regular harvest is to be found in the sports shooting interests in many temperate countries, where the numbers removed should be based on the above theory. In practice it is probable that many years' experience of the population, adjusted on a year-to-year basis to take account of changing conditions, e.g. severity of winter mortality and level of poaching, sets the limits to the cull.

Economic and social factors

The social benefits of game cropping have frequently emphasized the provision of meat in protein-deficient regions, and the generation of an economic return to the cropper, whether this be government or private individual. It is however important to recognize that increased provision of meat does not necessarily result in a corresponding increase in revenue. This has been illustrated by a computer simulation of the effects of three cropping strategies on female Himalayan thar (*Hemitragus jemlahicus*) in New Zealand. The three strategies were designed to maximize numbers, carcase weight and value as sustained yields. Account has to be taken of various economic factors. For example, the value of skins declines although the value of meat increases with age. In difficult terrain, a small animal may be transported by the hunter and a large one may require a helicopter to remove it. The net profit is higher for a smaller carcass as overheads are reduced. The analysis (Table 8.2) shows that to obtain MSY for numbers the youngest animals are those to take, while in financial return it is best to settle for a smaller annual take in numbers but to select the slightly older individuals. To produce an MSY of meat is relatively expensive. It is also noteworthy that an unselective age-group cropping for MSY of numbers produces a good yield of meat and an intermediate financial return. This is of practical value as the selective removal of a particular age group is difficult in the field.

The economics of African game cropping is poorly documented, with most accounts indicating no more than the market value of the meat produced. An exception is the accounts of the Doddieburn-Manyoli Ranch in Zimbabwe where detailed expenditure attributable to the ranching was balanced against the income from sales of hides and meat (Mossman and Mossman, 1976). Of the available accounts between 1960 and 1969, the range showed a profit in four years out of eight. In two years of losses, foot-and-mouth restrictions halted cropping for parts of the

Table 8.2 Computer simulation of maximization of yield from female thar (after Caughley, 1977)

Yield maximal in terms of	Age group harvested	Relative yield		
		nos.	*kg*	$
Numbers	0–1	100	900	200
Weight	3+	52	2095	52
Value	1–2	72	1800	720
Unselective	all	100	2062	398

years. Over the whole period there was a profit of 8.2% of the total revenue. On many farms practising game ranching, this is an activity additional to the main one of cattle rearing. It is also difficult to make meaningful comparisons of the economics of commercial game ranching with more conventional agriculture, as the latter frequently attracts government subsidy in various ways. Further problems are evident with respect to the financial viability of game cropping. It cannot be divorced from other forms of economic return from wildlife, notably tourism. A conflict of interests would arise between reducing the population to a level to produce an MSY and the viewing requirement of large numbers of animals. There can also be market fickleness as to the acceptability of game meat which can depress its price. Game cropping still requires the more general application of rigorous scientific method. When this has been applied for a protracted period its economic and social merits could be better assessed.

8.2 The conservation and management of whales

There are ten species of commercial whale. Nine belong to the filter-feeding baleen whales or Mysticeti, and one, the sperm whale (*Physeter macrocephalus*), is a toothed whale whose principal food is squid and fish. These animals range in size from the large blue whale (*Balaenoptera musculus*), up to 30 m long, to the small minke whale (*B. acutorostrata*) which is only 9 m. The remaining species are the right (*Eubalaena glacialis*), gray (*Eschrichtius robustus*), bowhead (*Balaena mysticetus*), humpback (*Megaptera novaeangliae*), fin (*Balaenoptera physalus*), sei (*B. borealis*) and Bryde's (*B. edeni*). All are worldwide in their distribution except the bowhead which is restricted to the Arctic seas.

While the whaling industry goes back many hundreds of years, it was the result of two technological advances that facilitated a much more effective exploitation of whale stocks. The first, in 1868, was the invention of the harpoon gun, making killing more efficient. The second was the introduction of factory ships in 1925. This made possible the taking of whales in oceans remote from shore stations. The exploitation of the southern oceans and their vast and previously untouched whale stocks commenced about 1905 but it was not until factory ships arrived that massive removal began. From 1925 it gradually increased up to the years immediately before the war when 46 000 whales were taken (Fig. 8.2). There was a decline in whaling in the war years and thereafter from 1946 to 1965 approximately 30 000 whales were removed each year.

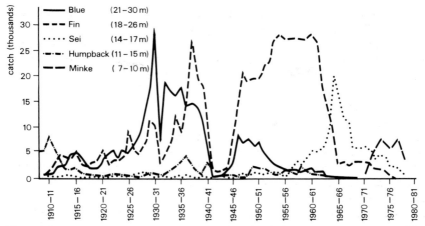

Figure 8.2 Catches of baleen whales in the southern hemisphere, 1910–1977. Approximate lengths (m) of adults of each species are indicated (modified from Allen, 1980).

Subsequently, more active conservation and depletion of stocks resulted in an annual removal of 10 000–15 000 animals. Species have not been taken in the same proportions throughout this period. The largest blue whales were the first priority and as their stocks declined the next largest, the fin whale, was taken and so until the latest whale, the minke, to attract attention is also the smallest. The sudden drop (Fig. 8.2) in the number of sei whale caught is the result of restrictive catch limits. The small catch of the humpback can probably be attributed to their presence in low numbers for a long time. Even so, their numbers fell dramatically in the early 1960s. A similar pattern of depletion to that in the southern oceans is also to be found in the lower-stocked North Pacific. Here an active whaling industry did not operate before the 1950s.

The most important commercial product of whales is oil. The baleen whales produce oil used in margarine and other foodstuffs; that produced from sperm whales is chemically different and used for lubricants. In addition the flesh of baleen whales has become important in recent years as meat for human consumption, notably in Japan. At present, only two nations, Japan and Russia, take the oceanic or pelagic whales although a further eight countries are involved in coastal whaling. The estimated weight of pelagic Antarctic whales removed fell from 3 m tons in 1938 to 660 000 tons in 1973. In 1938 whales amounted to 15% of the world's marine harvest; in 1980 the estimate was 1%. The drop can be attributed to increase in the take of fish as well as decline of the whale catch.

Table 8.3   Estimates of original and current whale populations ($\times 1000$) in the southern hemisphere. (Original figures are estimates at commencement of whaling operations.) After Allen, 1980

	Sperm		Right	Hump-back	Blue	Fin	Sei	Bryde's	Minke
	♂	♀							
Original	625	625	100+	130	220	490	191	30	205
Current	460	510	3	3	11	103	37	30	205

Estimates have been obtained of whale populations at present and at the commencement of whaling. Examination of the southern hemisphere numbers (Table 8.3) reveals considerable differences in the levels of exploitation. Lightly exploited species are Bryde's, minke and female sperm and are in contrast to the severely depleted right, humpback, blue, fin and sei. Obtaining estimates of this type is notoriously difficult and several of these figures must be subject to large margins of error. Yet the determination of harvesting levels must be based on a knowledge of whale stocks. Since 1948, international whaling has been regulated by the International Whaling Commission, one of whose objectives is to ensure that whale stocks are maintained at a level "... for orderly development of the whaling industry". Inevitably, this introduces the concept of sustained and maximum sustained yields. The powers of the IWC are wide-ranging and include the designation of protected species or stocks, the setting of catch limits by species, areas and possibly sexes, the limitation of areas where factory ships can be used, the inspection of whaling practices and the collection of scientific data. The species now protected are right, bowhead, gray, blue and humpback.

In 1975 IWC adopted a new management strategy which involved dividing whale stocks into three categories, viz. (i) initial management stocks capable of having their numbers reduced to a level of produce MSY (to be in this category the stocks must be at a level more than 20% above that to produce MSY); (ii) sustained management stocks maintained as MSY; and (iii) protection stocks whose levels are below the sustained management level and require full protection. For harvesting, quotas are set at 90% of MSY. No quota is set on unexploited initial management stocks until an estimate of stock size has been obtained. If the yield from a sustained management stock has remained unaltered for several seasons without any evidence of the stock declining, the quota can be set at the mean level of the previous catches.

Estimation of whale stocks is so difficult that recourse is made to several methods and sources of information accumulated over several years. One estimate of numbers can be obtained by relating effort and catch (Delury method) providing the population is not static (Gambell, 1976). By plotting effort against yield it is possible to extrapolate the relationship to zero effort, this being the unexploited population level. Here the standardization of effort is important, as there is a constant trend to improvement of efficiency in methods of location and capture. Allowance must also be made for changes in recruitment (reproductive rate) that may occur within the population at different densities. This is available from animals caught, as also are changes in age structure. The combination of catch and effort statistics with age composition data can provide total mortality estimates. Marking techniques, using stainless steel darts, can be used in mark-recapture estimates. Allowance has to be made for darts not obtaining attachment and unevenness of recovery and reporting. A useful measure from marking, not affected by poor reporting, is the percentage of returns made in successive years from one year's marking. The resulting decline provides a measure of mortality.

Population decline has had several noticeable effects on the reproductive biology of the survivors. Over thirty years the blue and fin whales have increased their pregnancy rates in the southern hemisphere from about 25 to 50% of the adult females pregnant, while the sei whale has shown an increase from 25 to 60%. Age of attainment of sexual maturity often falls in lower populations. In the southern hemisphere fin whale it fell from 9–10 years in 1930 to 6 in 1950.

In economic terms the objectives must be to produce the greatest yield at the lowest unit cost. If the effort, and therefore the cost, is great and the return small the venture is in danger of becoming unviable. As populations approach low densities, the effort can become too large to produce a satisfactory return. In setting quotas, the IWC aims to have a permitted catch 90% of MSY. This is satisfactory providing MSY can be accurately estimated. Underestimation would produce a lower sustained yield while overestimation will result in population decline if its effects pass unnoticed. The manifestations of a declining population would be higher reproductive rates, changing age structure of the population and greater effort required to sustain the yield.

Other management procedures have been suggested for IWC protected and yielding stocks. Some years ago, a ten-year moratorium was proposed on all whale species. Various difficulties lie in its implementation, including the unwillingness of certain nations to accede to the proposal. In any event, it is argued, no greater protection would be afforded to those species already protected while the industry would be deprived of those species from which a regular yield can be obtained. Cessation of whaling would also prevent information being obtained on the effects of the present worldwide management of whale stocks. A suggested alternative approach involved the implementation of different management strategies on comparable stocks. This would provide information on density-dependent population characters. To obtain this data, some stocks would probably have to be reduced to low levels; a practice unattractive in several quarters. Furthermore, the time taken to produce useful results would take several years. While this scheme would not reduce overall catches, the application of different levels of removal for different stocks could lead to political difficulties of acceptability.

8.3 Management of African elephants

The geographical range of the African elephant (*Loxodonta africana*) used to be more extensive than at present. This contraction can be attributed to increase in the human population, resulting in urbanization, modification of habitat, exclusion from many localities, as well as more effective methods of illegal pursuit and killing for ivory. Today elephants are found in savanna and forests in thinly populated and remote parts of the continent, and in reserves and parks where some measure of protection is usually conferred upon them. Management strategies and methods within conserved areas have been the subject of considerable debate, partly as a result of our limited knowledge of their ecology and partly due to more emotive and less rational issues.

The elephant population of Tsavo Park, Kenya, has been subject to a number of pressures since the mid-1950s, which become evident as the subsequent course of population change is traced (Cobb, 1980). The interaction between elephant and habitat and the effects of man (directly and indirectly) all pose problems for those endeavouring to maintain a balanced healthy population. Many of the events recorded in Tsavo are paralleled elsewhere in Africa with their resolution in this locality equally applicable in other places. Tsavo illustrates many of the general problems of elephant management.

Tsavo Park is situated in the dense *Acacia* and bush savanna with an annual rainfall fluctuating widely about a mean of approximately 500 mm. The Park itself covers 21 000 km^2, although adjacent to it is further suitable elephant habitat, bringing the total to 45 000 km^2. The larger area is referred to as the Tsavo Ecological Unit. The Park was gazetted in 1948 but it was not until 1957 that anti-poaching operations were effective.

Population estimates, within the Park, in 1962 were between 6800 and 10 800 elephant. Over the next few years there was a rapid increase in numbers to over 15 000 in 1965 and 23 000 in 1967 (Laws, 1969*b*). By 1970 the total within the Ecological Unit was put at 42 000. This is a relatively dry area of low productivity for so large and rapidly increasing a population, which was probably accounted for to some extent by immigration. By this time these high populations were removing bush and destroying trees and converting the vegetation to grassland (Laws, 1970). 1970 and 1971 were years of exceptionally low rainfall, and as a result available water, which elephant require daily, was of limited and restricted distribution. Elephant congregated about such water as was available and fed intensively on the vegetation in its vicinity. It was the Eastern sector of the Park where the severest effects of the drought were experienced. The strength of the mother–calf bond was such as to limit foraging by the adult cows to the maximum distance the calf could travel each day from the water. Inevitably with these high populations food ran out and starvation followed. As the bulls had no attached young they could forage further and enjoyed higher survival than the cows and calves. During these two years an estimated 5900 elephant died out of 20 000 living in the Eastern sector (Corfield, 1973). However, the total within the whole Ecological Unit was probably around 36 000 until 1975. Then came the poachers who between 1975 and 1979 killed an estimated 24 000 elephant. This leaves 12 000 elephant within the whole area; of these 6000 are probably within the Park. This is a remarkable account of unnaturally rapid expansion, consequent habitat deterioration and a dramatic decline at a time of climatic adversity to be followed by an even more dramatic illegal removal.

The rapid growth of the population in the 1960s posed the question, should the population be culled? It has been suggested that severe droughts as occurred in 1970 and 1971 appear every 50 years in Tsavo and that there is then a sudden drop in the elephant population (Phillipson, 1975). This takes time to recover and while this is happening the trees can regenerate for 30–35 years in the absence of heavy browsing pressure. There would thus be a stable limit cycle which would result in a cyclical balance between the numbers of trees and elephant. The proponents of this cycle would not recommend culling. Unfortunately, there is little evidence of a 50-year climatic cycle at Tsavo, and, perhaps more significantly, trees do not form a major component of the diet. An alternative hypothesis is that elephants have been displaced from their equilibrium (possibly due to immigration) with their food supply and adjustments, i.e. culling, are necessary to redress the balance (Malpas, 1978). Elephants range over large areas and it has been suggested that under natural conditions they may exploit a favoured area intensively for a number of years before moving to an alternative location. This cycle of movement would cover large areas and take place over many years. In this way, elephant would be in equilibrium with the resources of the total area, even though at one time a part of it may be subjected to heavy utilization and the rest may be uninhabited. Decades may elapse between visits to a particular locality. Under present man-modified conditions this could mean that elephants are "trapped" within a Park, with nowhere available to move on to, and be in excess of the permanently supportable population within so small an area. Whatever the causes of these high population densities, there seems little justification for permitting elephants to adjust their populations through self-regulation.

Elephants, like whales, have various intraspecific methods of regulating their populations (Laws, 1969a). These include an increase in the mean calving interval (section 3.4), and in the age at first pregnancy, from 12 years in low to 18 years in high populations. Important as these mechanisms may be, they are too slow to be effective regulators in current conditions. There is also evidence that the Tsavo elephants have years of high and low recruitment, which probably correspond to times of favourable and unfavourable resource availability, manifesting themselves in differing conception rates and infant survival.

The managerial problem with poachers is somewhat different as scientific judgements are less involved. Here there are two main courses of action. The first is prevention, and to accomplish this, more intensive patrolling is necessary; the second involves making the product (ivory) unmarketable. Bans on export and sales within the country are essential.

8.4 Feral cats

The feral cat is descended from the domestic cat (*Felis catus*). It differs from the latter in breeding in the wild and living independently of man. There are degrees of independence, with its extent frequently depending on the habitat in which the cat is living. Rural feral cats in areas where human populations are low frequently adopt similar feeding habits to wild cats by consuming vertebrate and insect prey. Many farms, some with colonies of up to 30 or more animals, support what are probably semi-feral cats which partially depend on man for food but have no close social association. In urban situations, while living independent of man, feral cats frequently depend on garbage and handouts for their food.

Urban cats occur in many situations. In a recent survey (Rees, 1981), hospitals (65 %) figured prominently, followed by industrial sites (9 %) and private housing developments (5 %). It is probable that hospitals were over-represented in the sample; here cats live in outbuildings, cellars and inaccessible places and are not present in the wards. Other sites include docks, shopping centres, military establishments, power stations and educational institutions. Small colonies are more common than large ones: 44 % include 1–10 animals, 29 % 11–20, 20 % 21–50, and the remainder over 50. Most colonies are well established with 75 % in existence for more than 5 years and a further 10 % for 2 to 5 years. Socially, there are conflicting views on the benefits and disadvantages of the cats. Among the proposed advantages are those of keeping down pests, the interest provided for hospital patients and a feeling of companionship, albeit not as close as with the domestic cat. Cats were disliked because of their general nuisance, smells from faeces and urine, the presence of dead animals and the transfer of their fleas. The importance of many of these problems still needs more critical evaluation.

A detailed study of the population ecology of feral cats has been undertaken at Portsmouth Dockyard (Dards, 1979, 1981). This covers an area of 85 ha and is surrounded on the landward side by a wall 3–5 m high and elsewhere by dry docks and the sea. Within the dockyard are numerous buildings including workshops, offices, welding shops, boiler houses and canteens; there are also fenced storage compounds and transformers. In various parts of the dockyard are piles of wood, metal, boxes and crates, rudders, propellers and other items. There is a network of steam pipes radiating in underground ducts from the boiler houses. Most of the yard is surfaced by tarmac and has a considerable movement of motor vehicles in it each day. For the cats this is an ideal habitat for cover and warmth, particularly the ducts containing the steam pipes. Food comes from two sources. Large skips are temporarily placed alongside ships after docking, into which waste food is deposited. The second, and quantitatively more important, source is from dockyard workers. This is usually commercially produced cat food and, with milk, is often provided in excess. Other foods are incidental such as the occasional fish caught by fishermen, insects and the occasional rodent. Interestingly, the faeces contain large quantities of hair, presumably as a result of intensive grooming and considerable amounts of grass. The latter is thinly dispersed in the dockyards and must be actively searched by the cats. It is probably an emetic.

The population is relatively stable (Fig. 8.3) at the high density of

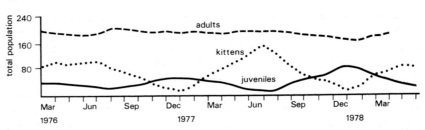

Figure 8.3 Population changes in feral cats in Portsmouth Dockyard, 1976–1978 (modified from Dards, 1979).

around 200 adults. The females have mean home ranges of 0.84 ha and live in social groups having an intensively-used core area. Twenty groups of 2 to 11 adults (average 4.5) live in the dockyard. The inclusion of kittens and juveniles swells the size of the groups. The amount of the increase varies seasonally. For example, in June 1976 the average group size was 10.5 (range 2–25). The members of groups display considerable amicable behaviour to each other with no high-intensity agonistic behaviour recorded. Adult males are solitary, have an average range size of 8.4 ha, are more mobile than females and often encompass several female groups within one of their ranges. The time a male spends with each group is variable and not necessarily equally apportioned between them. Male ranges overlap, but as a result of their relatively large size encounters are few.

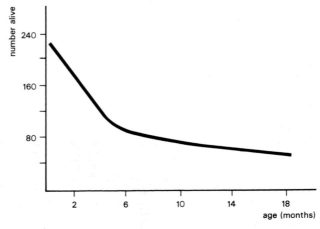

Figure 8.4 Survivorship of feral cats in Portsmouth Dockyard (modified from Dards, 1979).

Most births occur between March and September, although they can occur in any month of the year (Fig. 8.3). Kitten mortality is high, mainly as a result of the cat 'flu complex of viruses and feline panleucopaenia (feline infectious enteritis). Early mortality of kittens can account for the loss of 30% of births (Fig. 8.4) and over the first year the total can be reduced by two-thirds. Of the adults (i.e. animals more than one year) nearly half are under 3 years old. One of the major causes of adult death is car strike. A noteworthy feature is the low productivity of females, which may be socially induced. Certain females have more pregnancies than others, the same females being more prolific in successive years. There is also a significant negative correlation between the number of pregnancies per female and the average number of females in a group (Fig. 8.5). The females in smaller groups have larger numbers of pregnancies. There is, however, no correlation between the number of females in a group and the number of kittens produced. How social behaviour regulates numbers is not clear. It may, for example, inhibit conception or stimulate early abortion. What is particularly interesting is the apparent self-regulatory population limitation in a situation of excess food supply.

The extent to which these dockyard feral cats require control is a matter of value judgement. That they are fed, generally well cared for and reasonably healthy as adults, suggest that their presence is valued and they experience as comfortable a life as many domestic cats, although it must be recognized that there is a high kitten mortality. Reduction of the population could be effected by not putting out food, although denying

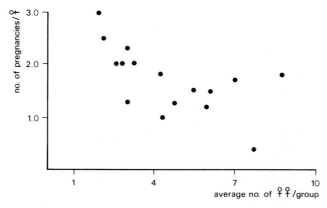

Figure 8.5 Mean numbers of pregnancies in female feral cats in relation to group size. There is a significant negative correlation, $r = -0.62$ (modified from Dards, 1979).

access to the skips would be more difficult. Various techniques have been devised for catching cats and a number of strategies put forward for reducing colony size, e.g. neutering and release, regular control culling, and application of chemical birth control.

8.5 Rabies and fox ecology

The rabies virus can occur in many species of mammal, particularly carnivores, but also including man, bats, cattle and sheep. There are several methods of transmission, ranging from bites from infected animals and consumption of infected food to transplantation of infected corneas. The virus incubates from two to eight weeks before manifestation of the symptoms. If the virus establishes itself in the brain, the furious symptoms of terror of water (hydrophobia), muscular spasms of the diaphragm and other respiratory muscles, violent retching and vocal modification result. Infection of the spinal cord produces the "dumb" symptoms of paralysis. Treatment by vaccine after exposure but prior to manifestation of the symptoms has a high chance of success. This is contrary to the situation after the symptoms have appeared when there is little hope of recovery.

Rabies spreads in waves. Currently Europe is experiencing such a wave which started at Gdansk in Poland in 1939 (Macdonald, 1980). It has moved across 1600 km of Europe, reaching central France in 1978, and is still advancing. There are various types of rabies. This is the sylvatic rabies in which foxes (*Vulpes vulpes*) account for 74% of the recorded cases. Susceptibility is high in other canids and some rodents and lower in domestic cats, cattle, horses and goats. Elsewhere in Europe, notably to the south-east, canine rabies is the endemic type. This is most common in domestic dogs and cats and farm animals. Its incidence in wild carnivores is extremely low. The range, extension, and establishment of sylvatic rabies has three phases. There is the epizootic phase as it enters a new area (Fig. 8.6) during which there is high fox mortality. From 40 km to 100 km behind this front there comes a zone of low fox populations where disease transmission is limited and the presence of rabies less evident. Under these conditions, fox populations recover and a further outbreak then occurs as in the enzootic phase (Fig. 8.6). Meanwhile the front is moving on. The silent phase often gives the false impression that the disease has swept through the area and is no longer present. This is not the case. Throughout its range there is a serious health hazard to man as well as considerable economic consequences through cattle deaths. In France, considerable protection has been afforded to domestic animals through vaccination.

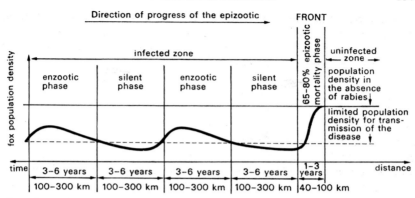

Figure 8.6 Fluctuations in fox density as a rabies epizootic moves through an area (modified from Macdonald, 1980).

The rate and pattern of spread of sylvatic rabies must be linked to fox behaviour, as one animal passes on the infection to another. The incidence of infection in foxes varies seasonally, being low in June and July and high in March; males have a higher infection rate in winter and females in late spring. The mobility of fox cubs varies with sex and age. For the first few months after birth, movements are generally small but by December 73% of males and 34% of females have moved more than 5 km from the parental den. The journeys made by some individuals can be very considerable; distances of 257 km and 394 km have been recorded in Denmark from marking of cubs to recapture. The most itinerant animals are sub-adult males.

Foxes are shy, elusive animals of nocturnal habits. In Britain they live in rural and urban situations. The latter has been studied in Oxford where extended family groups of up to six adults have been observed to defend territories. Here itinerant foxes are rejected, although one animal was observed to visit five of the six territories in Oxford in one night. Mean territory size here was 44.8 ha. In a similar study in Bristol (Harris, 1980), home range (it is uncertain whether the ranges are defended) size of adult vixens was 45.4 ha with many of the ranges overlapping. These animals would make occasional sorties of considerable distances outside their ranges.

Rural areas support much lower densities, with home range estimates of 40–100 ha (Lloyd, 1980) in west Wales, to 1300 ha in the upland sheep farming areas of northern England. Comparison of high and low density populations in rural Wales indicates much greater distances moved by young males in the latter where conditions are less favourable. Thus fox population densities show considerable differences, and while there is always appreciable movement, the frequency of contact between individuals (important in the context of disease transmission) must vary in relation to these different conditions.

Like most mammals the fox has considerable reproductive flexibility. In the coniferous forests of northern Sweden, mean litter size is low in years of low rodent numbers. This considerably influences the population demography from year to year. Where food supplies are less variable in the agricultural southern Sweden, mean litter size and population demography show greater stability. Social factors may also influence the number of adult females that breed. Thus the depression of a population through rabies could release several reproductive inhibitors and maximize production.

One approach to rabies control involves ascertaining the densities to which foxes have to be reduced to ensure that contact is so infrequent that the disease does not spread. Several mathematical models have been put forward which endeavour to incorporate the most important parameters. These models still require testing against real situations. In parts of Britain, fox populations are relatively high, and it is probable that in the event of sylvatic rabies reaching this country local fox populations would have to be dramatically reduced. The concept of reducing numbers to an acceptable minimal level has attractions as it is extremely difficult to completely eliminate an elusive and ubiquitous species like the fox.

A vigorous and successful fox reduction campaign was undertaken in southern Denmark as infected foxes entered from Germany. Here foxes were killed by gassing and strychnine nitrate poisoning in infected areas and in a belt 20 km north of the epizootic front. Bounty shooting of foxes also took place but this was less effective in reducing numbers. A less intensive campaign was mounted in a further 20 km strip across the country. The vaccination of domestic dogs was compulsory. The control operation was initiated in 1964 and was successful in preventing the spread of rabies until the pressure was released in 1969 when an outbreak of rabies occurred. The release of pressure again in 1977 had a similar result. It is important to recognize the resilience of the disease in re-establishing itself within a depressed fox population. One of the side effects of fox removal was a substantial increase in the numbers of hare, partridge and pheasant.

The vaccination or immunization of foxes by self-administered drugs poses numerous problems. Firstly the fox must be prepared to take the bait containing the vaccine. The most effective vaccines (ERA strains of BHK 21) must be absorbed in the tissues of the mouth and not the stomach. The vaccines are unstable; they deteriorate in a few hours at temperatures above 4°C. And finally, vaccines that produce immunity in foxes may produce a mild form of disease if consumed by other species; the consequences of this are considerable. This research is still in its infancy—to ensure the target animal takes the bait and the vaccine then acts efficiently are demanding tasks. The control of rabies may ultimately depend on a combination of methods of which the control of the fox population would be one.

8.6 Conservation of otters in England

Until, and for some time after, the Second World War, the otter (*Lutra lutra*) was widespread throughout the lakes and river systems of England. Being a wide ranging carnivorous species its numbers were probably never high by comparison with, for example, herbivorous species of the same or slightly smaller size. Nevertheless, it was a ubiquitous and not infrequently seen member of the fauna. While occasionally found on land the otter has an appreciable dependence on water, where it spends much of its time feeding when not lying up in its holt. In Sweden (Erlinge, 1968), individual otters cover extensive areas the precise size depending on availability of water, topography, food supply and population density. Males have a mean range diameter of 15 km and females with young 7 km. Otters can also travel considerable distances overnight with up to 10 km recorded for males. The occurrence of otters is frequently recorded from their droppings or spraints which are often deposited in prominent positions along the banks of watercourses.

To determine changes in numbers, recourse has to be made to the records of otter hunts (Chanin and Jefferies, 1978). Nine hunts were active in England between 1950 and 1976, each with its defined hunting area. Two were recorded in the north (Kendal, Northern Counties), three in the middle of the country (Wye Valley, Buckinghamshire, Eastern Counties) and four in the south (Dartmoor, Culmstock, Courtney Tracy, Crowhurst). The hunts kept records of otters killed and what are called "finds" which are occasions when they are seen by huntsmen. The interpretation of data from this source must be treated with some caution as hunts may tend, when numbers are low, to direct their activities to locations where they know or believe otters are present. In this way comparability between years can be biased. Even so, when allowance is made for this factor the data from the hunts are of considerable value. The record of finds for the Culmstock Otterhounds (Fig. 8.7), whose country

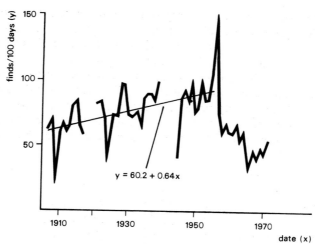

Figure 8.7 Records of finds by the Culmstock Otter Hounds 1907–1971. The regression line is fitted for the period 1907–1955 (modified from Chanin and Jefferies, 1978).

extends over parts of Devon, Somerset and Dorset shows a fluctuating and steady increase from 1907 to 1955 and an appreciable drop from 1956 to 1966. A fall over this period was recorded nationally, although its magnitude varied from hunt to hunt, being comparatively small in the two northern packs. The major decline started in 1957 and 1958. The hunts responded to lower numbers by adopting a policy in 1964 of restricting kills, with the result that by 1976 only 7% of finds were killed. This compares to a mean of around 50% for the Culmstock Hunt in the interwar years. From 1978 the otter was given full legal protection.

From 1977 to 1979 a national otter survey was conducted under the sponsorship of the Nature Conservancy Council (Lenton *et al.*, 1980). The impracticability of surveying the whole country necessitated using a sampling technique. This was achieved by examining alternate quarters of each 100 km square of the national grid, and within each, selecting 600 m strips of bank at 5–8 km distances along all water courses including rivers, canals, lake shores and coastlines. Highly urbanized areas were omitted. Of 2940 sites surveyed, otters were found at only 170 or 6%. (The comparable figures for Wales and Scotland are 20 and 73% respectively). The otter is sparse or absent from most of central England. It was most frequently recorded from the south-west, Welsh borders, parts of East Anglia and the north-east. It is also present in other parts of the north and south. In many of these latter areas otters are apparently only present as small, localized, remnant populations and must therefore be considered highly vulnerable (Fig. 8.8). The current picture is thus one of generally low numbers, and even where they are relatively well established, as in the south-west, their future viability cannot go unquestioned.

How has this decline come about? Has it halted? What measures can be taken to give the otter its previous status? The reasons for the decline are not known. Numbers dropped, and inevitably few dead animals were recovered to ascertain cause of death. Nevertheless, some information is available and from our knowledge of otter ecology and physiology it is

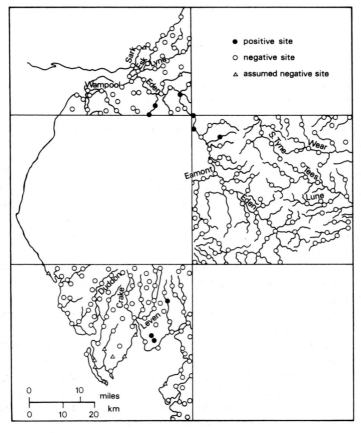

Figure 8.8 Distribution of signs of otters in north-west England 1977–79. The empty 50 km squares were not surveyed (modified from Lenton *et al.*, 1980).

possible to infer some of the contributory factors that have effected loss of numbers. Removal of cover, disturbance and pollution are probably of greatest significance. The clearing of waterside vegetation (whether this be trees, undergrowth or emergent plants) deprives otters of adequate cover in which to lie up. Hollows often develop in the vicinity of tree roots, particularly ash, which provide suitable holts. The tidying up of these habitats by riparian landowners has been to the otter's disadvantage. Otters prefer to utilize a stretch of water without interference by man. Disturbance is or has been caused by bankside fishing, boating and hunting. The increasing use of canals, lakes and rivers by motor-powered boats has steadily reduced favourable habitat. Otters avoid situations where there is temporary bankside residence as with caravans and camping. Providing the intensity of an activity is not too great, e.g. limited fishing by day, and plenty of cover is available, otters may remain, but a level of interference is reached when they will vacate an area completely. Clearance and disturbance could not in themselves account for the sharp decline in the 1950s. The drop at this time could be correlated with the then widespread use of the organochlorine insecticide dieldrin. This commenced in the mid-1950s when dieldrin was used as a sheep dip and for cereal and sugar beet dressings. It has a high fat solubility and a persistence which facilitate transfer along the food chain to species such as the otter. High levels have been recorded in herons, which, like the otter, are fish feeders. The extensive use of dieldrin was stopped by voluntary restriction in the 1960s and 1970s. Even so over 80 % of otters examined post-mortem between 1963 and 1973 contained measurable quantities and in the occasional animal it was present at a lethal level. While fresh water quality has been improved in recent years, there was at least until 1978 evidence of dieldrin in a number of river systems.

England and Welsh national hunting statistics indicate an appreciable drop in the average annual number of otter killed (205) from 1950 to 1957 compared to pre-war (c. 400 +). Furthermore, the numbers killed and the percentages of finds killed varied considerably from pack to pack. Courtney Tracy killed 165 (58 % of finds) and Northern Counties 50 (38 % of finds). The northern hunts, where the reported decline was less severe, were exerting lower removal. From 1958 to 1963 the 11 active hunts in England and Wales killed 1065 otters and while this represents a fall to 178 per year, it was a high removal at a time when numbers were dropping. Other causes such as severe winters, disease, and trapping and killing, may have reduced numbers but in themselves they have probably not been sufficient to effect a substantial and sustained decline.

The status of the otter can be improved in several ways. Firstly, legal protection has been afforded by designating it an endangered species so that killing and taking are now illegal. More positively, habitat improvements could be implemented (Macdonald *et al.*, 1978). These include the provision of patches of trees and dense scrub at regular, approximately 1 km, intervals along river banks to provide lying-up sites. The insertion of lengths of ceramic piping can also perform a similar function, while limitation of recreation and provision of clean water systems are also extremely important.

8.7 Rodents and human disease

Diseases and infections naturally transmitted between vertebrate animals and man are called zoonoses. Many are known to occur in rodents (Table 8.4), although not all seriously affect human health. Thus, while rodents can carry anthrax and brucellosis, these diseases hardly ever infect man. In contrast, the effects of a disease like plague on human populations can be devastating (Arata, 1975) and the role of rats in its transmission extremely important.

Table 8.4   Some human diseases found in rodents. From Cox, 1979.

Disease in man	Causative organism	Transmission
Bacterial zoonoses		
Tularemia	*Francisella*	Blood sucking arthropods
Salmonellosis	*Salmonella*	Contamination of food
Leptospirosis	*Leptospira*	Urine contaminating water
Plague	*Yersinia*	Flea vectors; droplet infection
Rickettsial zoonoses		
Kenya typhus	*Rickettsia*	Tick larva
Q fever	*Coxiella*	Tick
Scrub typhus	*Rickettsia*	Tick
Viral zoonoses		
Lymphatic choriomeningitis	Arenavirus	Inhalation of infective particles
Lassa fever		Ingestion of excreta
Bolivian haemorrhagic fever	Arenavirus	Ingestion of excreta
Far Eastern encephalitis	Flavivirus	Tick
Colorado tick fever	Orbivirus	Tick
Rift Valley fever		Mosquito
Protozoal zoonoses		
Leishmaniasis	*Leishmania*	Sandfly
Toxoplasmosis	*Toxoplasma*	Consumption of undercooked meat

Under natural conditions the infectious agents, the host, and the vector exist together as part of the ecosystem (Cox, 1979). The three organisms have a set of ecological requirements such as climatic, edaphic and vegetation conditions, and live together in a balanced relationship. Some localities are more favourable than others for the maintenance of the infection and it is these that form the permanent or elementary foci of infection. Foci can be either restricted, with clearly delineated boundaries, or diffuse, with the margins less apparent. They are often in situations uninhabited by man so that transmission does not take place. Infection occurs when man moves into these areas. The occasional visitor, e.g. camper, tourist, soldier, forester, may briefly encounter the focus, while activities involving changing land use, e.g. forest clearance, agricultural development, or urbanization, can bring man into a closer, more permanent association with the focus. Furthermore, modification of the habitat can create a more beneficial environment to the host. An example is the forest clearance of Eastern Europe and Central Asia for pasture and meadows, which has increased the habitat suitable for the field vole (*Microtus*). In this region the vole is an important reservoir of the tularemia bacterium. Bolivean haemorrhagic fever virus occurs in the cricetine rodent, *Calomys*. This animal is commensal with man, particularly in depressed urban conditions.

The situation is more complex in the transmission of plague in South Africa, where man has inadvertently created in his farms and buildings suitable habitats for the multimammate mouse (*Praomys*), which is not however the primary host of plague. The main reservoir is the highveld gerbil, *Tatera brantsi*, with transmission through its flea *Xenopsylla philoxera*. This gerbil has little or no contact with man. There is an intricate relationship between gerbil densities and intensity of infection which results in an irregularity of frequency of plague outbreaks. In an experimental area in the highveld, outbreaks occurred in 1940, 1948 and 1954 before the bacterium was eliminated. When an outbreak of plague occurs, many gerbils die and *Praomys* moves into the empty burrows. This species has a wider ecological range than the gerbil and part of its range extends to human habitations. Plague then spreads through this population and into close association with man. Spread can be accelerated through the house rat (*Rattus*) population with which the multimammate mouse comes into contact. This sequence oversimplifies the actual course of events as more than one species of flea is involved. However, it illustrates the ecological complexity of the transmission route. Plague is widespread throughout the world and over 200 species of rodent are known to harbour the bacterium.

These are from all the major families including squirrels, rats and mice, gerbils and cricetines. With such a variety of species there are correspondingly numerous ecological routes of transmission. The situation is further complicated because some species are more resistant to plague than others.

The emergence of new diseases is constantly being recorded. Lassa fever was first reported from West Africa in 1970. Even though it still has localized foci, there were early records considerable distances apart in Nigeria and Sierra Leone. Other recent discoveries include Kyasanur forest disease (1957), Bolivian haemorrhagic fever (1959) and Marburg virus (1967). While control of zoonoses is meeting with modest success, complete eradication remains a remote possibility. There are various methods of control and prevention. One approach is immunization against and the treatment of human subjects with the disease but in the present context, interest lies more in prevention through control of rodents. In urban areas this can be effected by intensive use of traps, rodenticides and repellants. Additionally, cleaner, more hygienic habits, adequate disposal of waste food and the rodent-proofing of buildings minimize the likelihood of establishment of rodent populations. Such measures are often relatively expensive and therefore not easy to implement in many developing and poorer countries.

Rodents in field situations are notoriously difficult to control. Substantial reductions, e.g. by poisoning, can result in reduced rates of infection between individuals. Alternatively, attempts at habitat modification, e.g. removal of weed patches and scrub in agricultural land, can create less suitable conditions and thereby suppress populations. There are often remarkably fundamental gaps in our knowledge of rodent ecology which prevent the implementation of control measures, particularly as many diseases associated with rodents occur in the developing world. Here information is often minimal or absent on such aspects as foods, habitat requirements, home ranges, movements and population dynamics of the indigenous species. The situation can sometimes be further complicated by problems of identification and imprecision in species diagnosis. These problems of field study and control are frequently exacerbated by poor assessment of the economic and public health impact of rodents.

BIBLIOGRAPHY

More general works are marked with an asterisk *

Abaturov, B. D. and Kuknetosov, G. V. (1976) Formation of secondary biological production by little sousliks. *Zool. Zhuru* **55**, 1526–1537.

*Allen, K. R. (1980) *Conservation and Management of Whales*. University of Washington Press, Seattle.

Altman, P. L. and Dittmer, D. S. (1972) *Biological Data Book*. I. 2nd edn., American Societies for Experimental Biology, Bethesda.

Anderson, P. K. (1961) Density, social structure, and nonsocial environment in house-mouse populations and the implications for regulation of numbers. *Trans. New York Acad. Sci.*, 2nd ser., **23**, 447–451.

Arata, A. A. (1975) The importance of small mammals in public health, in *Small Mammals: their Productivity and Population Dynamics*, Golley, F. B., Petrusewicz, K. and Ryszkowski, L. (eds.) Cambridge University Press, pp. 349–359.

*Asdell, S. A. (1964) *Patterns of Mammalian Reproduction*, 2nd edn., Cornell University Press, Ithaca.

Ayeni, J. S. O. (1972) Utilisation of waterholes in Tsavo National Park (East). *E. Afr. Wildl. J.* **13**, 305–324.

Baker, J. R. (1930) The breeding season in British wild mice. *Proc. zool. Soc. Lond.* **1930**, 113–126.

*Baker, R. R. (1978) *The Evolutionary Ecology of Animal Migration*. Hodder & Stoughton, London.

Barkalow, F. S. Jr., Hamilton, R. B. and Soots, R. F. Jr. (1970) The vital statistics of an unexploited gray squirrel population. *J. Wildl. Mgmt.* **34**, 489–500.

Batzli, G. O. (1975) The role of small mammals in arctic ecosystems, in *Small Mammals: their Productivity and Population Dynamics*, Golley, F. B., Petrusewicz, K. and Ryszkowski, L. (eds.) Cambridge University Press, pp. 243–268.

Bertram, B. (1975) Social factors influencing reproduction in wild lions. *J. Zool., Lond.* **177**, 463–482.

Bland, K. P. (1973) Reproduction in the female African tree rat (*Grammomys surdaster*). *J. Zool., Lond.* **171**, 167–176.

Blaxter, K. L. and Graham, N. M. (1955) Plane of nutrition and starch equivalents. *J. agric. Sci.* **46**, 292–306.

Bradbury, J. W. (1977) Social organisation and communication, in *Biology of Bats* III., Wimsatt, W. A. (ed.) Academic Press, New York, pp. 1–72.

Brambell, F. W. Rogers (1935) Reproduction in the common shrew (*Sorex araneus* L.). *Phil. Trans.* B **225**, 1–62.

Brambell, F. W. R. and Rowlands, I. W. (1936) Reproduction of the bank vole (*Evotomys glareolus* Schreber). I. The oestrous cycle of the female. *Phil. Trans.* B **226**, 71–97.

Breymeyer, A. I. and Van Dyne, G. M. (eds.) (1978) *Grasslands, Systems Analysis and Man*. Cambridge University Press.

Brosset, A. (1968) La permutation du cycle sexuel saisonnier chez le Chiroptère *Hipposideros caffer* au voisinage de l'equateur. *Biol. Gabon.* **4**, 325–341.

Brown, L. E. (1956) Field experiments on the activity of the small mammals (*Apodemus, Clethrionomys* and *Microtus*). *Proc. zool. Soc. Lond.* **126**, 549–564.

Brown, L. E. (1969) Field experiments on the movements of *Apodemus sylvaticus* L. using trapping and tracking techniques. *Oecologia* **2**, 198–222.

Carpenter, C. R. (1940) A field study in Siam of the behavior and social relations of the gibbon (*Hylobates lar*). *Comp. Psychol. Monogr.* **16**, 1–212.

*Caughley, G. (1977) *Analysis of Vertebrate Populations*. Wiley, Chichester.

Chanin, P. R. F. and Jeffries, D. J. (1978) The decline of the otter *Lutra lutra* L. in Britain: an analysis of hunting records and discussion of causes. *Biol. J. Linn. Soc.* **10**, 305–328.

Chappell, M. A. (1978) Behavioral factors in the altitudinal zonation of chipmunks (*Eutamias*). *Ecology* **59**, 565–579.

Cheeseman, C. L. and Delany, M. J. (1979) The population dynamics of small rodents in a tropical African grassland. *J. Zool., Lond.* **188**, 451–475.

Clutton-Brock, T. H. and Harvey, P. H. (1977) Primate ecology and social organisation. *J. Zool., Lond.* **183**, 1–39.

Cobb, S. (1980) Tsavo. *Swara* **3(4)**, 12–17.

Coe, M. J., Cummings, D. H. and Phillipson, J. (1976) Biomass and production of large African herbivores in relation to rainfall and primary production. *Oecologia* **22**, 341–354.

Coetzee, C. G. (1965) The breeding season of the multimammate mouse *Praomys* (*Mastomys*) *natalensis* (A. Smith) in the Transvaal Highveld. *Zool. Afr.* **1**, 29–40.

Conaway, C. H., Baskett, T. S. and Toll, J. E. (1960) Embryo resorption in the swamp rabbit. *J. Wildl. Mgmt.* **24**, 197–202.

*Corbet, G. B. and Hill, J. E. (1980) *A World List of Mammalian Species*. British Museum (Nat. Hist.), London.

*Corbet, G. B. and Southern, H. N. (1977) *The Handbook of British Mammals*, 2nd edn., Blackwell, Oxford.

Corfield, T. F. (1973) Elephant mortality in Tsavo National Park, Kenya. *E. Afr. Wildl. J.* **11**, 339–368.

Coupland, R. T. (ed.) (1979) *Grassland Ecosystems of the World*, Cambridge University Press.

Coupland, R. T. and Van Dyne, G. M. (1979) Systems synthesis, in *Grassland Ecosystems of the World*, Coupland, R. T. (ed.) Cambridge University Press, pp. 97–106.

Cox, F. E. G. (1979) Ecological importance of small mammals as reservoirs of disease, in *Ecology of Small Mammals*, Stoddart, D. M. (ed.) Chapman & Hall, London, pp. 213–238.

Crowcroft, P. (1954) The daily cycle of activity in British shrews. *Proc. zool. Soc. Lond.* **123**, 715–729.

Crowcroft, W. P. and Rowe, F. P. (1963) Social organisation and territorial behaviour in the wild house mouse (*Mus musculus*). *Proc. zool. Soc. Lond.* **140**, 517–531.

Dards, J. (1979) *The Population Ecology of Feral Cats* (*Felis catus* L.) in Portsmouth Dock-yard. Ph.D. thesis, University of Bradford.

Dards, J. (1981) Habitat utilisation by feral cats in Portsmouth Dockyard, in *The Ecology and Control of Feral Cats*, Universities Federation for Animal Welfare, Potter's Bar, pp. 30–46.

Dasmann, R. F. (1964) *African Game Ranching*. Pergamon, London.

Dasmann, R. F. and Mossmann, S. L. (1961) Commercial use of game animals on a Rhodesian ranch. *Wildlife* **3(3)**, 7–14.

Davis, D. H. S. (1953) Plague in South Africa: a study of the epizootic cycle in gerbils (*Tatera brantsi*) in the northern Orange Free State. *J. Hygiene* **51**, 427–449.

Davis, W. H. and Hitchcock, H. B. (1965) Biology and migration of the bat *Myotis lucifugus* in New England. *J. Mammal.* **46**, 296–313.

Dawson, J. J. and Hulbert, A. J. (1970) Standard metabolism, body temperature and surface area of Australian marsupials. *Am. J. Physiol.* **218**, 1233–1238.

Deevey, E. S. (1947) Life tables for natural populations of animals. *Quart. Rev. Biol.* **22**, 283–314.

Delany, M. J. (1964) An ecological study of the small mammals in Queen Elizabeth Park, Uganda. *Revue Zool. Bot. afr.* **70**, 129–147.

Delany, M. J. (1971) The biology of small rodents in Mayanja Forest, Uganda. *J. Zool., Lond.* **165**, 85–129.

Delany, M. J. (1972) The ecology of small rodents in tropical Africa. *Mammal Rev.* **2**, 1–42.

Delany, M. J. and Bishop, I. R. (1960) The systematics, life history and evolution of the bank vole *Clethrionomys* Tilesius in north-west Scotland. *Proc. zool. Soc. Lond.* **35**, 409–422.

Delany, M. J. and Neal, B. R. (1969) Breeding seasons in rodents in Uganda. *J. Reprod. Fert., Suppl.* **6**, 229–236.

*Delany, M. J. and Happold, D. C. D. (1979) *Ecology of African Mammals.* Longman, London.

Dieterlen, F. (1967) Jahreszeiten und Fortpflanzungsperioden bei den Muriden des Kivusee-Gebietes (Congo). I. Ein Beitrag zum Problem der Populationsdynamik in Tropen. *Z. Säugetierk* **32**, 1–44.

Drozdz, A. (1979) Seasonal intake and digestibility of natural foods by roe-deer. *Acta theriol.* **24**, 137–170.

Dueser, R. D. and Shugart, H. H. (1979) Niche pattern in a forest-floor small-mammal fauna. *Ecology* **60**, 108–118.

Dunmire, W. W. (1960) An altitudinal survey of reproduction in *Peromyscus maniculatus. Ecology* **14**, 174–182.

Dutt, R. H. and Bush, L. F. (1955) The effect of low environmental temperature on initiation of the breeding season and fertility in sheep. *J. Anim. Sci.* **14**, 885–896.

Einarson, A. E. (1956) Life of the mule deer, in *The Deer of North America*, Taylor, W. P. (ed.) Stackpole, Harrisburg.

Eisenberg, J. F., Muckenhirn, N. A. and Rudran, R. (1972) The relation between ecology and social structure in primates. *Science* **176**, 863–874.

Ellefson, J. O. (1968) Territorial behavior in the common white-handed gibbon, *Hylobates lar* Linn, in *Primates: Studies in Adaptation and Variability*, Jay, P. C. (ed.), Holt, Rinehart & Winston, New York, pp. 180–199.

Ellenbroek, F. J. M. (1980) Interspecific competition in the shrews *Sorex araneus* and *Sorex minutus* (Soricidae, Insectivora): a population study of the Irish pygmy shrew. *J. Zool., Lond.* **192**, 119–136.

Elton, C. and Nicholson, M. (1942) The ten-year cycle of the lynx in Canada. *J. Anim. Ecol.* **11**, 215–244.

*Eltringham, S. K. (1979) *The Ecology and Conservation of Large African Mammals.* Macmillan, London.

Eltringham, S. K. (1980) A quantitative assessment of range usage by large African mammals with particular reference to the effects of elephants on trees. *Afr. J. Ecol.* **18**, 53–71.

Erlinge, S. (1968) Territoriality of the otter *Lutra lutra* L. *Oikos* **19**, 81–98.

*Ewer, R. F. (1973) *The Carnivores.* Weidenfeld & Nicolson, London.

Ferns, P. N. (1980) Energy flow through small mammal populations. *Mammal Rev.* **10**, 165–188.

Field, C. R. (1972) The food habits of wild ungulates in Uganda by analyses of stomach contents. *E. Afr. Wildl. J.* **10**, 17–42.

Field, C. R. (1974) Scientific utilization of wildlife for meat in East Africa: a review. *J. Sth. Afr. Wildl. Mgmt. Ass.* **4**, 177–183.

Field, C. R. (1979) Game ranching in Africa, in *Applied Biology* IV, Coaker, T. H. (ed.), Academic Press, London, pp. 63–101.

Field, C. R. and Blankenship, L. H. (1973a) Nutrition and reproduction of Grant's and Thomson's gazelles, Coke's hartebeest and giraffe in Kenya. *J. Reprod. Fert. Suppl.* **19**, 287–301.

Field, C. R. and Blankenship, L. H. (1973b) On making the game pay. *Africana* **5(4)**, 22–23.

Fitzgerald, B. M. (1977) Weasel predation on a cyclic population of the montane vole (*Microtus montanus*) in California. *J. Anim. Ecol.* **46**, 367–397.

Flake, L. D. (1973) Food habits of four species of rodents on a short-grass prairie in Colorado. *J. Mammal.* **54**, 636–647.

Fleming, T. H. (1970) *Artibeus jamaicensis*: delayed embryonic development in a Neotropical bat. *Science* **171**, 402–404.

Fleming, T. H. (1971) Population ecology of three species of Neotropical rodents. *Misc. Publ. Mus. Zool., Univ. Michigan* **143**, 1–77.

Fleming, T. H. (1973) Numbers of mammal species in North and Central American forest communities. *Ecology* **54**, 555–563.

Fleming, T. H. (1975) The role of small mammals in tropical ecosystems, in *Small Mammals: their Productivity and Population Dynamics*, Golley, F. B., Petrusewicz, K. and Ryszkowski, L. (eds.) Cambridge University Press, pp. 269–298.

Flowerdew, J. R. (1972) The effect of supplementary food on a population of wood mice (*Apodemus sylvaticus*). *J. Anim. Ecol.* **41**, 553–566.

Flux, J. E. C. (1967) Reproduction and body weights of the hare *Lepus europaeus* (Pallas) in New Zealand. *N. Z. Jl. Sci.* **10**, 357–401.

Flux, J. E. C. (1969) Current work· on the African hare, *Lepus capensis* L. in Kenya. *J. Reprod. Fert. Suppl.* **6**, 225–227.

Ford, J. (1971) *The Role of Trypanosomiasis in African Ecology: a Study of the Tsetse Problem.* Oxford University Press.

French, N. R., Grant, W. E., Grodzinski, W. and Swift, D. M. (1976) Small mammal energetics in grassland ecosystems. *Ecol. Monogr.* **46**, 201–220.

French, N. R., Stoddart, D. M. and Bobek, B. (1975) Patterns of demography in small mammal populations, in *Small Mammals: their Productivity and Population Dynamics*, Golley, F. B., Petrusewicz, K. and Ryszkowski, L. (eds.) Cambridge University Press, pp. 73–102.

Gaisler, J. (1979) Ecology of bats, in *Ecology of Small Mammals*, Stoddart, D. M. (ed.) Chapman & Hall, London, pp. 281–342.

Gambell, R. (1976) World whale stocks. *Mammal Rev.* **6**, 41–53.

Gebczynski, M. (1975) Heat economy and the energy cost of growth in the bank vole during the first months of postnatal life. *Acta theriol.* **20**, 379–434.

Ghobrial, C. I. and Hodieb, A. S. K. (1973) Climate and seasonal variations in the breeding of the desert jerboa *Jaculus jaculus* in the Sudan. *J. Reprod. Fert. Suppl.* **19**, 221–233.

Gibb, J. A. and Flux, J. E. C. (1978) Mammals, in *The Natural History of New Zealand*, Williams, G. R. (ed.) Reed, Wellington, pp. 334–371.

Glover, T. D. and Sale, J. B. (1968) The male reproductive tract in the rock hyrax. *J. Zool., Lond.* **156**, 351–361.

Godfrey, G. K. and Crowcroft, P. (1960) *The Life of the Mole.* Museum Press, London.

Golley, F. B. (1960) Energy dynamics of a food chain of an old-field community. *Ecol. Monogr.* **30**, 187–206.

*Golley, F. B. and Buechner, H. K. (eds.) (1968) *A Practical Guide to the Study of the Productivity of Large Herbivores.* Blackwell, Oxford.

*Golley, F. B., Petrusewicz, K. and Ryszkowski, L. (eds.) (1975) *Small Mammals: their Productivity and Population Dynamics.* Cambridge University Press.

Gorecki, A. and Grygielska, M. (1975) Consumption and utilization of natural foods by the common hamster. *Acta theriol.* **20**, 237–246.

Gosling, L. M. (1969) Parturition and related behaviour in Coke's Hartebeest *Alcelaphus buselaphus cokei* Gunther. *J. Reprod. Fert. Suppl.* **6**, 265–286.

Grant, P. R. (1972) Interspecific competition among rodents. *Ann. Rev. Ecol. System.* **3**, 79–106.

Grodzinski, W., Makomaska, M., Tertil, R. and Wiener, J. (1977) Bioenergetics and total impact in vole populations. *Oikos* **29**, 494–510.

Grodzinski, W. and Wunder, B. A. (1975) Ecological energetics of small mammals, in *Small Mammals: their Productivity and Population Dynamics*, Golley, F. B., Petrusewicz, K. and Ryszkowski, L. (eds.) Cambridge University Press, pp. 173–206.

Gromadzki, M. and Trojan, P. (1971) Estimation of population density in *Microtus arvalis* (Pall) by three different methods. *Ann. Zool. Fenn.* **8**, 54–59.

Gwynne, M. D. and Bell, R. H. V. (1968) Selection of vegetation components by grazing ungulates in the Serengeti National Park. *Nature* **220**, 390–393.

Haas, H. and Ricou, G. A. E. (1979) Consumers in meadows and pastures, in *Grassland Ecosystems of the World*, Coupland, R. T. (ed.) Cambridge University Press, pp. 139–153.

Hall, K. R. L. and De Vore, I. (1965) Baboon social behavior, in *Primate Behavior: Field Studies of Monkeys and Apes*, De Vore, I. (ed.) Holt, Rinehart & Winston, New York, pp. 53–110.

Happold, D. C. D. (1967) Guide to the natural history of Khartoum Province. III. Mammals. *Sudan Notes Rec.* **48**, 111–132.

Happold, D. C. D. (1977) A population study of small rodents in the tropical rainforest of Nigeria. *Terre Vie* **31**, 385–457.

Harestad, A. S. and Bunnell, E. L. (1979) Home range and body weight—a re-evaluation. *Ecology* **60**, 389–402.

Harris, S. (1980) Home ranges and patterns of distribution of foxes (*Vulpes vulpes*) in an urban area as revealed by radio tracking, in *Handbook on Biotelemetry and Radio Tracking*, Almaner, C. R. and Macdonald, D. W. (eds.) Pergamon, Oxford, pp. 685–690.

Harrison, J. L. (1955) Data on the reproduction of some Malayan mammals. *Proc. zool. Soc. Lond.* **125**, 445–460.

Harrison, J. L. (1962) The distribution of feeding habits among animals in a tropical rain forest. *J. Anim. Ecol.* **31**, 53–63.

Hayward, G. F. and Phillipson, J. (1979) Community structure and functional role of small mammals in ecosystems, in *Ecology of Small Mammals*, Stoddart, D. M. (ed.) Chapman & Hall, London, pp. 135–211.

Heptner, V. G. (1967) *Mammals of the Soviet Union*. II. Moscow (pp. 636–686 on stoat and weasel, English translation, British Library RTS 6458).

Hewer, H. R. (1974) *British Seals*. Collins, London.

Humphrey, S. R. (1974) Zoogeography of the nine-banded armadillo (*Dasypus novemcinctus*) in the United States. *Bioscience* **24**, 457–462.

Jarman, P. J. (1974) The social organisation of antelope in relation to their ecology. *Behaviour* **58**, 215–267.

Jarvis, J. U. M. (1981) Eusociality in a mammal: co-operative breeding in naked mole-rat colonies. *Science* **212**, 571–573.

Jarvis, J. U. M. and Sale, J. B. (1971) Burrowing and burrow patterns in East African mole rats *Tachyoryctes, Heliophobius* and *Heterocephalus*. *J. Zool., Lond.* **163**, 451–479.

Jenkins, D. and Harper, R. J. (1980) Ecology of otters in northern Scotland. II. Analyses of otter (*Lutra lutra*) and mink (*Mustela vison*) faeces from Deeside, N. E. Scotland in 1977–78. *J. Anim. Ecol.* **49**, 737–754.

Jewell, P. A. (1966) The concept of home range in mammals. *Symp. zool. Soc. Lond.* **163**, 451–479.

Jezierski, W. (1977) Longevity and mortality rate in a population of wild boar. *Acta theriol.* **24**, 337–348.

Jordan, P. A., Botkin, D. B. and Wolfe, M. L. (1971) Biomass dynamics in a moose population. *Ecology* **52**, 147–152.

Kaufman, D. W. and Kaufman, G. A. (1975) Caloric density of the old-field mouse during postnatal growth. *Acta theriol.* **20**, 83–95.

Kaufmann, J. H. (1962) Ecology and social behaviour of the coati, *Nasua narica*, on Barro Colorado Island, Panama. *Univ. California Publ. Zool.* **60**, 95–222.

Kaufmann, J. H. (1974) Social ethology of the whiptail wallaby, *Macropus parryi*, in north-eastern New South Wales. *Anim. Behav.* **22**, 281–369.

Keast, A. (1969) Evolution of mammals of southern continents. 7. Comparisons of the contemporary mammalian faunas of the southern continents. *Quart. Rev. Biol.* **44**, 121–167.

King, J. A. (1955) Social behavior, social organisation, and population dynamics in a black-tailed prairiedog town in the Black Hills of South Dakota. *Contr. Lab. Vert. Biol., Univ. Michigan* **67**, 1–123.

King, J. E. (1964) *Seals of the World*. British Museum (Nat. Hist.), London.

King, J. M. and Heath, B. R. (1975) Game domestication for animal production in Africa. *World Anim. Rev.* **16**, 23–30.

Kleiber, M. (1961) *The Fire of Life*. Wiley, New York.

Klingel, H. (1969) The social organisation and population ecology of the plains zebra (*Equus quagga*). *Zool. Afr.* **4**, 249–263.

Krasinski, Z. A. (1978) Dynamics and structure of the European bison population in the Bialowieza Primeval Forest. *Acta theriol.* **23**, 3–48.

Krebs, C. J. and Myers, J. H. (1974) Population cycles in small mammals. *Adv. Ecol. Res.* **8**, 267–399.

Kruuk, H. (1972) *The Spotted Hyena: a Study of Predation and Social Behavior*. University of Chicago Press.

Kruuk, H. and Turner, M. (1967) Comparative notes on predation by lion, leopard, cheetah and wild dog in the Serengeti area, East Africa. *Mammalia* **31**, 1–27.

Kühme, W. (1965) Communal food distribution and division of labour in African hunting dogs. *Nature* **205**, 443–444.

Kunz, T. H. (1973) Resource utilization: temporal and spatial components of bat activity. *J. Mammal.* **54**, 14–32.

Laurie, E. M. O. and Hill, J. E. (1954) *List of Land Mammals of New Guinea, Celebes and Adjacent Islands, 1758–1952*. British Museum (Nat. Hist.), London.

Lawick, H. van (1974) *Solo: the Story of an African Wild Dog*. Houghton Mifflin, Boston.

Lawick, H. van and Lawick-Goodall, J. van (1971) *Innocent Killers*. Houghton Mifflin, Boston.

Laws, R. M. (1968) Dentition and ageing of the hippopotamus. *E. Afr. Wildl. J.* **6**, 19–52.

Laws, R. M. (1969*a*) Aspects of reproduction in the African elephant, *Loxodonta africana*. *J. Reprod. Fert. Suppl.* **6**, 193–218.

Laws, R. M. (1969*b*) The Tsavo Research Project. *J. Reprod. Fert., Suppl.* **6**, 495–531.

Laws, R. M. (1970) Elephants as agents of habitat and landscape change in East Africa. *Oikos* **21**, 1–15.

Leatherwood, S. and Walker, W. A. (1979) The northern right whale dolphin *Lissodelphis borealis* Peale in the eastern north Pacific, in *Behavior of Marine Animals 3: Cetaceans*, Win, H. E. and Olla, B. L. (eds.) Plenum, New York, pp. 85–141.

Lecyk, M. (1962) The effect of the length of daylight on reproduction in the field vole *Microtus arvalis* (Pall.). *Zoologica Pol.* **12**, 189–221.

Lenton, E. J., Chanin, P. R. F. and Jefferies (1980) *Otter Survey of England 1977–79*. Nature Conservancy Council, London.

Lidicker, W. Z. (1975) The role of dispersal in the demography of small mammals, in *Small Mammals: their Productivity and Population Dynamics*, Golley, F. B., Petrusewicz, K. and Ryszkowski, L. (eds.) Cambridge University Press, pp. 103–128.

Lloyd, H. G. (1963) Intrauterine mortality in the wild rabbit (*Oryctolagus cuniculus* L.) in populations of low density. *J. Anim. Ecol.* **32**, 549–563.

Lloyd, H. G. (1964) *Influences of environmental factors on some aspects of breeding in the wild rabbit Orcytolagus cuniculus* (L.). M.Sc. thesis, University College of North Wales, Bangor.

Lloyd, H. G. (1970) Variation and adaptation in reproductive performance. *Symp. zool. Soc. Lond.* **26**, 165–187.

Lloyd, H. G. (1980) *The Red Fox*, Batsford, London.

Lord, R. D. (1961) Magnitudes of reproduction in cottontail rabbits. *J. Wildl. Mgmt.* **25**, 28–33.

Lowe, V. P. W. (1969) Population dynamics of the red deer (*Cervus elaphus* L.) on Rhum. *J. Anim. Ecol.* **38**, 425–457.

McConnell, B. R. and Dalke, P. D. (1960) The Cassia deer herd of southern Idaho. *J. Wildl. Mgmt.* **24**, 265–271.

*Macdonald, D. W. (1980) *Rabies and Wildlife*. Oxford University Press.

*McDonald, P., Edwards, R. A. and Greenhalgh, J. F. D. (1973) *Animal Nutrition*, 2nd edn., Oliver & Boyd, Edinburgh.

Macdonald, S. M., Mason, C. F. and Coghill, I. S. (1978) The otter and its conservation in the River Teme catchment. *J. Appl. Ecol.* **15**, 373–384.

McNab, B. K. (1979) The influence of body size on the energetics and distribution of fossorial and burrowing mammals. *Ecology* **60**, 1010–1021.

McNeil, S. and Lawton, J. H. (1970) Animal production and respiration in animal populations. *Nature* **225**, 472–474.

MacKinnon, K. S. (1978) Stratification and feeding differences among Malayan squirrels. *Malay. Nat. J.* **30**, 593–608.

MacPherson, T. H. (1964) A northward range extension of the red fox in the eastern Canadian arctic. *J. Mammal.* **45**, 138–140.

Malpas, R. C. (1978) *The Ecology of the African Elephant in Rwenzori and Kabalega Falls National Parks, Uganda*. Ph.D. thesis, University of Cambridge.

*Maynard, L. A., Loosli, J. K., Hintz, H. F. and Warner, R. G. (1979) *Animal Nutrition*, 7th edn., McGraw-Hill, New York.

Mead, R. A. (1968a) Reproduction in western forms of the spotted skunk (genus *Spilogale*). *J. Mammal.* **49**, 373–390.

Mead, R. A. (1968b) Reproduction in eastern forms of the spotted skunk (genus *Spilogale*). *J. Zool.* **156**, 119–136.

Mech, L. D. (1966) *The Wolves of Isle Royale*. U.S. Nat. Park Serv., Fauna Ser. 7.

Mech, L. D. (1970) *The Wolf: the Ecology and Behavior of an Endangered Species*. Natural History Press, Garden City.

Misonne, X. (1963) Les rongeurs du Ruwenzori et des regions voisines. *Explor. Parc. Natn. Deux. Ser.* **14**, 1–164.

*Mitchell, B., Staines, B. W. and Welch, D. (1977), *Ecology of Red Deer. A Research Review Relevant to their Management in Scotland*. Institute of Terrestrial Ecology, Cambridge.

*Moen, A. N. (1973) *Wildlife Ecology*. Freeman, San Francisco.

Montgomery, W. I. (1980) Population structure and dynamics of sympatric *Apodemus* species (Rodentia: Muridae). *J. Zool., Lond.* **192**, 351–377.

Moor, P. P. and Steffens, F. E. (1972) The movements of vervet monkeys (*Cercopithecus aethiops*) within their home range as revealed by radio-tracking. *J. Anim. Ecol.* **41**, 677–687.

Mossman, S. L. and Mossman, A. S. (1976) *Wildlife Utilization and Game Ranching*. Occ. Paper no. 17, International Union for the Conservation of Nature, Morges.

Mutere, F. A. (1967) The breeding biology of equatorial vertebrates: reproduction in the fruit bat *Eidolon helvum*, at latitude 0°21′N. *J. Zool., Lond.* **153**, 153–161.

Myers, K. (1966) The effects of density on sociality and health in mammals. *Proc. ecol. Soc. Aust.* **1**, 40–64.

Myllymäki, A. (1975) Control of field rodents, in *Small Mammals: their Productivity and Population Dynamics*, Golley, F. B., Petrusewicz, K. and Ryszkowski, L. (eds.) Cambridge University Press, pp. 311–338.

Myrcha, A. (1975) Bioenergetics of an experimental population and individual laboratory mice. *Acta theriol.* **20**, 175–226.

Naumov, N. P. (1975) The role of rodents in ecosystems of the northern deserts of Eurasia, in *Small Mammals: their Productivity and Population Dynamics*, Golley, F. B., Petrusewicz, K. and Ryszkowski, L. (eds.) Cambridge University Press, pp. 299–310.

Neal, B. R. (1970) The habitat distribution and activity of a rodent population in western Uganda, with particular reference to the effects of burning. *Revue Zool. Bot. Afr.* **81**, 29–50.

Neal, B. R. (1977) Reproduction of the punctated grass-mouse, *Lemniscomys striatus* in the Ruwenzori National Park, Uganda (Rodentia: Muridae). *Zool. Afr.* **12**, 419–428.

Neal, B. R. (1981) Reproductive biology of the unstriped grass rat, *Arvicanthis* in East Africa. *Z. Säugetierk.* **46**, 174–189.

Neal, E. G. and Harrison, R. J. (1958) Reproduction in the European badger (*Meles meles* L.). *Trans zool. Soc. Lond.* **29**, 67–131.

Nel, J. A. J. (1978) Habitat heterogeneity and changes in small mammal community structure and resource utilization in the southern Kalahari. *Bull. Carnegie Mus. Nat. Hist.* **6**, 118–131.

O'Roke, E. C. and Hamerstrom, F. N. (1948) Productivity and yield of the George Reserve deer herd. *J. Wildl. Mgmt.* **12**, 78.

Pearson, O. P. (1964) Carnivore-mouse predation: an example of its intensity and bioenergetics. *J. Mammal.* **45**, 177–188.

Pearson, O. P. (1966) The prey of carnivores during one cycle of mouse abundance. *J. Mammal.* **40**, 169–180.

Pearson, O. P. (1971) Additional measurements of the impact of carnivores on California voles (*Microtus californicus*). *J. Mammal.* **52**, 41–49.

Pennycuick, L. (1975) Movements of the migratory wildebeest population in the Serengeti area between 1960 and 1973. *E. Afr. Wildl. J.* **13**, 65–87.

Pernetta, J. C. (1976) Bioenergetics of British shrews in grassland. *Acta theriol.* **21**, 481–498.

Petrides, G. A. and Swank, W. G. (1965) Population densities and range-carrying capacity for large mammals in Queen Elizabeth National Park, Uganda. *Zool. Afr.* **1**, 209–225.

Petrides, G. A. and Swank, W. G. (1966) Estimating the productivity and energy relations of an African elephant population. *Proc. Ninth Int. Grass. Cong. Sao Paulo, Brazil*, 831–842.

Phillipson, J. (1975) Rainfall, primary production and 'carrying capacity' of Tsavo National Park (East) Kenya. *E. Afr. Wildl. J.* **13**, 171–201.

Poczopko, P. (1979) Metabolic rate and body size relationships in adult and growing homeotherms. *Acta theriol.* **24**, 125–136.

Racey, P. A. (1973) The viability of spermatozoa after prolonged storage by male and female European bats. *Per. Biologorum (Zagreb)* **75**, 201–205.

Rahm, U. (1970) Note sur la reproduction des Sciuridés et Muridés dans la forêt equatoriale au Congo. *Rev. Suisse Zool.* **77**, 635–646.

Rees, P. (1981) The ecological distribution of feral cats and the effects of neutering on a hospital colony, in *The Ecology and Control of Feral Cats*, Universities Federation for Animal Welfare, Potter's Bar, pp. 12–22.

Ricker, W. E. (1958) *Handbook of Computations for Biological Statistics of Fish Populations*. Fis. Res. Bd. Canada Bull. no. 199.

Robinette, W. L. and Child, G. F. T. (1964) Notes on biology of the lechwe (*Kobus leche*). *Puku* **2**, 84–117.

Rood, J. P. (1975) Population dynamics and food habits in the banded mongoose. *E. Afr. Wildl. J.* **13**, 89–112.

*Sadleir, R. M. F. S. (1969) *The Ecology of Reproduction in Wild and Domestic Mammals.* Methuen, London.

Schaller, G. B. (1970) This gentle and elegant cat. *Nat. Hist.* **79**, 31–39.

Schaller, G. B. (1972) *The Serengeti Lion: a Study in Predator–Prey Relations.* University of Chicago Press.

Scheffer, V. B. (1951) The rise and fall of a reindeer herd. *Sci. Monthly* **73**, 356–362.

Schreiber, R. K. (1978) Bioenergetics of the Great Basin pocket mouse *Perognathus parvus. Acta theriol.* **23**, 469–487.

Schreiber, R. K. and Johnson, D. R. (1975) Seasonal changes in body composition and caloric content of Great Basin rodents. *Acta theriol.* **20**, 343–364.

Scott, J. A., French, N. R. and Leetham, J. W. (1979) Patterns of consumption in grasslands, in *Perspectives in Grassland Ecology*, French, N. R. (ed.) Springer-Verlag, pp. 89–105.

Sharman, G. B. (1970) Reproductive physiology of marsupials. *Science* **167**, 1221–1228.

Sharman, G. B. and Berger, P. J. (1969) Embryonic diapause in marsupials. *Adv. Reprod. Physiol.* **4**, 211–240..

Sheppe, W. and Osborne, T. O. (1971) Patterns of use of a flood plain by Zambian mammals. *Ecol. Monogr.* **41**, 179–205.

Shield, J. W. and Woolley, P. (1963) Population aspects of delayed birth in the quokka (*Setonix brachyurus*). *Proc. zool. Soc. Lond.* **141**, 783–790.

Short, R. V. and Hay, M. F. (1966) Delayed implantation in the roe deer *Capreolus capreolus. Symp. zool. Soc. Lond.* **15**, 173–194.

Simpson, G. G. (1964) Species density of North American recent mammals. *Syst. Zool.* **13**, 57–73.

Sinclair, A. R. E. (1974) The natural regulation of buffalo populations in East Africa. IV. The food supply as a regulating factor, and competition. *E. Afr. Wildl. J.* **12**, 291–311.

Sinclair, A. R. E. (1975) Resource limitations in tropical grasslands. *J. Anim. Ecol.* **44**, 497–520.

Sinclair, A. R. E. (1977) *The African Buffalo. A Study in Resource Limitation of Populations.* University of Chicago Press.

Slijper, E. J. (1962) *Whales.* Hutchinson, London.

Simpson, G. G. (1965) *The Geography of Evolution.* Chilton Books, Philadelphia.

Smith, M. H. and McGinnis, J. T. (1968) Relationships of latitude, altitude and body size to litter size and mean annual production of offspring in *Peromyscus. Res. Pop. Ecol.* **10**, 115–126.

Smith, W. J., Smith, S. L., Oppenheimer, E. C., De Villa, J. G. and Ulmer, F. A. (1973) Behavior of a captive population of black-tailed prairie dogs: annual cycle of social behavior. *Behavior.* **46**, 189–220.

Smythe, M. (1966) Winter breeding in woodland mice, *Apodemus sylvaticus* and voles, *Clethrionomys glareolus* and *Microtus agrestis* near Oxford. *J. Anim. Ecol.* **35**, 471–485.

Smythe, N. (1970) The adaptive value of the social organisation of the coati (*Nasua narica*). *J. Mammal.* **51**, 818–820.

Southern, H. N. (1979a) Population processes in small mammals, in *Ecology of Small Mammals*, Stoddart, D. M. (ed.) Chapman & Hall, London, pp. 63–102.

Southern, H. N. (1979b) The stability and instability of small mammal populations, in *Ecology of Small Mammals*, Stoddart, D. M. (ed.) Chapman & Hall, London, pp. 103–134.

Southwick, C. H. (1958) Population characteristics of house mice living in English corn ricks: density relationships. *Proc. zool. Soc. Lond.* **131**, 163–175.

Spence, D. H. N. and Angus, A. (1971) African grassland management—burning and grazing in Murchison Falls National Park Uganda, in *The Scientific Management of Animal and Plant Communities for Conservation*, Duffey, E. and Watt, A. S. (eds.) Blackwell, Oxford, pp. 319–332.

*Stoddart, D. M. (ed.) (1979) *Ecology of Small Mammals.* Chapman & Hall, London.

Svarc, S. S., Bol'sakov, V. N., Olenev, V. G. and Pjastolova, O. A. (1969) Population dynamics of rodents from northern and mountainous geographical zones, in *Energy Flow Through Small Mammal Populations*, Petrusewicz, K. and Ryszkowski, L. (eds.) Polish Scientific Publishers, Warsaw, pp. 205–220.

Sviridenko, P. A. (1940) Pitanie myshevidnykh gryzunov i znachenie ikh v problemlevozobnovleniya lesa. *Zool. Zhurnal* **19**, 680–703.

Taylor, K. D. (1968) An outbreak of rats in agricultural areas of Kenya in 1962. *E. Afr. agric. for. J.* **34**, 66–77.

Taylor, K. D. and Green, M. G. (1976) The influence of rainfall on diet and reproduction in four African rodent species. *J. Zool., Lond.* **180**, 367–389.

Thompson, H. V. and Worden, A. N. (1956) *The Rabbit*. Collins, London.

Treus, V. and Kravchenko, D. (1968) Methods of rearing and economic utilisation of eland in Askaniya—Neva Zoological Park. *Symp. zool. Soc. Lond.* **21**, 395–411.

*Tyndale-Biscoe, H. (1973) *Life of Marsupials*. Edward Arnold, London.

*Vaughan, T. A. (1978) *Mammalogy*, 2nd edn., Saunders, Philadelphia.

Verme, L. J. (1965) Reproduction studies on penned white-tailed deer. *J. Wildl. Mgmt.* **29**, 74–79.

Vogel, P. (1976) Energy consumption of European and African shrews. *Acta theriol.* **21**, 195–206.

Volf, J. (1963) Einige Bemerkungen zur Aufzucht von Eisbären (*Thalarctos maritimus*) in Gefangenschaft. *Zool. Gart. Lpz.* **28**, 97–108.

*Wallace, A. R. (1876) *The Geographical Distribution of Animals*. Macmillan, London.

Watson, R. M. (1969) Reproduction of wildebeest, *Connochaetes taurinus albojubatus* Thomas, in the Serengeti Region and its significance to conservation. *J. Reprod. Fert. Suppl.* **6**, 287–310.

Watts, C. H. S. (1969) The regulation of wood mouse (*Apodemus sylvaticus*) numbers in Wytham Woods, Berkshire. *J. Anim. Ecol.* **38**, 285–304.

*Wilson, E. O. (1975) *Sociobiology*. Harvard University Press, Cambridge.

Wilson, V. J. and Roth, H. H. (1967) The effects of tsetse control operations on common duiker in Eastern Zambia. *E. Afr. Wildl. J.* **5**, 53–64.

Wimsatt, W. A. (1945) Notes on breeding behavior, pregnancy, and parturition in some vespertilionid bats of the eastern United States. *J. Mammal.* **26**, 23–33.

Wood, D. H. (1970) An ecological study of *Antechinus stuartii* (Marsupialia) in a south-east Queensland rain forest. *Austr. J. Zool.* **18**, 185–207.

Wood, J. E. (1958) Age structure and productivity in a gray fox population. *J. Mammal.* **39**, 74–86.

Woolley, P. (1966) Reproduction in *Antechinus* spp. and other dasyurid marsupials. *Symp. zool. Soc. Lond.* **15**, 281–294.

Wyatt, J. E. and Eltringham, S. K. (1974) The daily activity of the elephant in the Rwenzori National Park, Uganda. *E. Afr. Wildl. J.* **12**, 273–290.

*Yalden, D. W. and Morris. P. A. (1975) *The Lives of Bats*. David & Charles, Newton Abbot.

Zejda, J. (1966) Litter size in *Clethrionomys glareolus* Schreber 1780. *Zool. Listy* **15**, 193–206.

Periodicals

There are numerous scientific journals and periodicals in which papers on mammal ecology appear. Those listed below are international in their content; in addition, there are several journals that restrict their interests to a particular region. Those marked with an asterisk do not publish exclusively on mammalogy.

*Acta Theriologica, *Animal Behaviour, *Behaviour, *Ecology, *Ecological Monographs, Folia Primatologia, *Journal of Animal Ecology, *Journal of Applied Ecology, Journal of Mammalogy, Journal of Primatology, *Journal of Wildlife Management, *Journal of Zoology, *La Terre et Vie, Mammal Review, Mammalia, Zeitschrift fur Säugetierkunde, *Zeitschrift fur Tierpsychologie.*

Index

Those species and higher taxonomic ranks included in the index contain a substantive account in the text. There are also within the volume numerous brief references, in the text and tables, to species, genera and other taxa of mammals which are omitted from the index as their inclusion would make it unduly extensive. However, the subject headings within the index lead the reader to these groups, the principles they illustrate and their position in mammal taxonomy.